Direct Numerical Simulation of
Very-Large-Scale Motions in
Turbulent Pipe Flow

DIRECT NUMERICAL SIMULATION OF VERY-LARGE-SCALE MOTIONS IN TURBULENT PIPE FLOW

DISSERTATION

zur Erlangung des akademischen Grades
Doktoringenieur (Dr.-Ing.)

vorgelegt der
Fakultär für Maschinenbau der
Technischen Universität Ilmenau
von Herrn

Christian Bauer
geboren am 25.02.1984
in Köln

Erstgutachter:	Prof. Dr.-Ing. Claus Wagner
Zweitgutachter:	Prof. Dr. Jörg Schumacher
Drittgutachter:	Prof. Dr. Markus Uhlmann

Eingereicht am:	2. Juli 2020
Verteidigt am:	8. Dezember 2020

Bibliografische Information der Deutschen Nationalbibliothek
Die Deutsche Nationalbibliothek verzeichnet diese Publikation in der Deutschen Nationalbibliografie; detaillierte bibliografische Daten sind im Internet über http://dnb.d-nb.de abrufbar.
1. Aufl. - Göttingen: Cuvillier, 2021
Zugl.: (TU) Ilmenau, Univ., Diss., 2020

Nonnenstieg 8, 37075 Göttingen
Telefon: 0551-54724-0
Telefax: 0551-54724-21
www.cuvillier.de

1. Auflage, 2021
Gedruckt auf umweltfreundlichem, säurefreiem Papier aus nachhaltiger Forstwirtschaft.

ISBN 978-3-7369-7369-5
eISBN 978-3-7369-6369-6

برای مهسا

Erklärung

Ich versichere, dass ich die vorliegende Arbeit ohne unzulässige Hilfe Dritter und ohne Benutzung anderer als der angegebenen Hilfsmittel angefertigt habe. Die aus anderen Quellen direkt oder indirekt übernommenen Daten und Konzepte sind unter Angabe der Quelle gekennzeichnet.

Bei der Auswahl und Auswertung folgenden Materials haben mir die nachstehend aufgeführten Personen in der jeweils beschriebenen Weise ~~entgeltlich~~/unentgeltlich[1] geholfen:

1. Prof. Claus Wagner hat mich in seiner Rolle als Doktorvater und fachlicher Betreuer dieser Arbeit im üblichen Umfang durch themenbezogene Diskussionen unterstützt.

2. Die Arbeit wurde von meiner Kollegin Annika Köhne im Hinblick auf Rechtschreibung, Grammatik und Form korrekturgelesen.

Weitere Personen waren an der inhaltlich-materiellen Erstellung der vorliegenden Arbeit nicht beteiligt. Insbesondere habe ich hierfür nicht die entgeltliche Hilfe von Vermittlungs- bzw. Beratungsdiensten (Promotionsberater oder anderer Personen) in Anspruch genommen. Niemand hat von mir unmittelbar oder mittelbar geldwerte Leistungen für Arbeiten erhalten, die im Zusammenhang mit dem Inhalt der vorgelegten Dissertation stehen.

Die Arbeit wurde bisher weder im In- noch im Ausland in gleicher oder ähnlicher Form einer Prüfungsbehörde vorgelegt.

Ich bin darauf hingewiesen worden, dass die Unrichtigkeit der vorstehenden Erklärung als Täuschungsversuch bewertet wird und gemäß § 7 Abs. 10 der Promotionsordnung den Abbruch des Promotionsverfahrens zur Folge hat.

Göttingen, 15. Juni 2020

Christian Bauer

---

[1] Unzutreffendes bitte streichen

Acknowledgements

The computational resources provided by DLR on the high performance computing clusters SCART and CARA and by LRZ on SuperMUC under grant no. pr62zu are gratefully acknowledged.

I wish to express my heartfelt gratitude to my supervisor Prof. Claus Wagner for his guidance, his valuable counsel, and his support and for the elbow room he granted me during the past four years. The accomplishment of this thesis in its present form would not have been possible without him. I am also deeply indebted to Prof. Markus Uhlmann not only for being a referee of my thesis, but also for introducing me to the field of turbulence research. He has been an initial inspiration for my doctoral studies and my scientific career — and as we know from chaos: small changes in initial conditions can lead to drastically different results. Moreover, I would like to sincerely thank Prof. Jörg Schumacher for his role as a referee of my thesis. His fruitful comments helped further improve this document. In addition, I would like to express my appreciation to Prof. Fröhlich and Prof. Cierpka for being members of my committee and my Rigorosum examiners. I am also grateful to Prof. Bergmann for accepting to chair my PhD committee. I particularly wish to express my gratitude to Tiago Pestana, not only for being an old friend but also for sharing with me his enthusiasm in turbulence research as well as to Yoshi Sakai for his friendship and his support and encouragement as the supervisor of my Master's thesis. I would also like to thank Annika Köhne, who proofread this thesis and many other manuscripts. My genuine appreciation goes to my former and present colleagues, particularly Agathe, Alexander, Alexandra, Amin, Andre, Christian, Daniel, Daniela, Doris, Emir, Felix, Gerrit, Henning, Keith, Konstantin, Lars, Max, Michael, Philipp, Tim, Tobi, and Yulia for the great time we have shared. Furthermore, I am truly grateful to my family — to my lovely mother Elke above all, as well as to my sister Tine, my brother Thomas, and my father Henry — for their continuous and unconditional support.

Finally yet most importantly, I want to thank you, Mahsa, my love, my soulmate, my strange attractor, for being by my side. Your love, your support, and your opinion mean everything to me.

Related Publications

The content of this thesis has in parts been published in the following references. The contributions of the co-authors are highly acknowledged.

1. Bauer, C., Feldmann, D., and Wagner, C., On the convergence and scaling of high-order statistical moments in turbulent pipe flow using direct numerical simulations, *Physics of Fluids*, 29(12), 125105 (2017)

2. Bauer, C., Feldmann, D., and Wagner, C., Revisiting the higher-order statistical moments in turbulent pipe flow using direct numerical simulations, *New Results in Numerical and Experimental Fluid Mechanics XI*, vol. 136 in Notes on Numerical Fluid Mechanics and Multidisciplinary Design, p. 75–84, Springer International Publishing (2018)

3. Feldmann, D., Bauer, C., and Wagner, C., Computational domain length and Reynolds number effects on large-scale coherent motions in turbulent pipe flow, *Journal of Turbulence*, 19(3), p. 274–295 (2018)

4. Bauer, C., Von Kameke, A., and Wagner, C., Kinetic energy budget of the largest scales in turbulent pipe flows, *Physical Review Fluids*, 4(6), 064607 (2019)

5. Bauer, C. and Wagner, C., Scaling of high-order statistics in turbulent pipe flow, *Direct and Large-Eddy Simulation XI*, vol. 25 in ERCOFTAC Series, p. 537–543, Springer International Publishing (2019)

6. Bauer, C. and Wagner, C., Analysis of the energy budget of the largest scales in turbulent pipe flow, *Progress in Turbulence VIII*, vol 226 in Springer Proceedings in Physics, p. 113–118, Springer International Publishing (2019)

Abstract

Turbulent pipe flow is not only of importance to engineering applications but also of fundamental interest to the study of wall-bounded turbulence. In the latter field, besides the well-explored near-wall cycle of small-scale motions, the so-called very-large-scale motions (VLSMs) have been discovered to contribute to the fluid turbulence of high Reynolds numbers in the vicinity of solid walls. Arising from the outer-flow region, these motions interact with the much smaller scales close to the wall, and thus, are reflected in both the instantaneous picture of turbulence and the turbulence statistics. Although a number of studies aiming at the characterisation of VLSMs have been carried out, the processes that govern their interactions with the near-wall, small scales are still widely unknown. Accordingly, this research studies VLSMs in turbulent pipe flow by means of direct numerical simulations with friction Reynolds numbers up to $\mathrm{Re}_\tau = 2880$.

After establishing $L \geq 42R$ as the numerical domain length required to reliably capture VLSMs, the convergence and the scaling of different order moments of the velocity distribution are studied and discussed with respect to the underlying coherent structures. In addition to the well-known dependence of the streamwise Reynolds stress peak on the Reynolds number, the streamwise velocity skewness and the wall-normal velocity flatness very close to the wall are also detected to logarithmically depend on the Reynolds number. These dependencies are revealed to be caused by the interaction of the outer-flow VLSMs with the near-wall region. While the low-speed VLSMs are rather passive, the high-speed VLSMs are mainly responsible for the scale interactions.

The subsequent analysis of the streamwise energy budget equation of the filtered velocity field reveals that VLSMs obtain their energy from the mean velocity field via a production mechanism similar to the one known from the near-wall cycle. Compared to the small-scale motions in the near-wall cycle, no mean backscattering of energy from smaller scales into VLSMs is detected. Scales smaller than VLSMs are, however, involved in the spatial transport of the energy of VLSMs away from its production peak towards the wall and towards the bulk flow region. Although not spotted in statistical averages, an inverse energy cascade is tracked in the instantaneous ejecting low-speed VLSMs, whereas high-speed sweeping VLSMs feature the classical forward energy cascade. This behaviour is in contrast with that of the small-scale, near-wall cycle, where the high-speed sweeps correlate with the backscattering and the low-speed ejections with the forward scattering of energy. In both cases, the inter-scale energy flux related to sweeps is more intense. At any rate, the backscattering phenomenon mainly occurs in regions of high turbulent kinetic energy production.

In a word, the research sheds new light on our understanding of the interaction between VLSMs and the near-wall cycle and leads to a better grasp of turbulent pipe flow in general.

Kurzfassung

Die turbulente Rohrströmung ist nicht nur von Bedeutung für Anwendungen im Ingenieurbereich, sondern auch für die Grundlagenforschung von wandnaher Turbulenz. Auf letzterem Gebiet wurden, neben dem gut untersuchten Wandzyklus kleinskaliger turbulenter Schwankungen, in jüngerer Zeit sogenannte „sehr großskalige Strömungsbewegungen" (engl.: very-large-scale motions, VLSMs) entdeckt. Letztere treten bei hohen Reynoldszahlen auf und interagieren mit den kleinskaligen Strömungsstrukturen in der Nähe der Wand. Obwohl sich eine Vielzahl von Studien mit der Charakterisierung von VLSMs beschäftigt, sind die Prozesse, welche deren Interaktion mit kleinskaligen Strukturen beherrschen weitgehend unbekannt. Die vorliegende Arbeit widmet sich daher dem Studium von VLSMs in der turbulenten Rohrströmung mittels direkter numerischer Simulation bei Schubspannungs-Reynoldszahlen bis $\mathrm{Re}_\tau = 2880$.

Nachdem die virtuelle Rohrlänge, welche benötigt wird, damit VLSMs realistisch simuliert werden können, auf $L \geq 42R$ festgelegt wurde, wird das Konvergenz- und Skalierungsverhalten verschiedener statistischer Momente der Geschwindigkeitsverteilung untersucht und in Bezug auf turbulente kohärente Strukturen diskutiert. Neben der Abhängigkeit des Maximums der axialen Reynolds'schen Spannung von der Reynoldszahl, sind die logarithmischen Reynoldszahlabhängigkeiten der Schiefe der axialen Geschwindigkeitsverteilung sowie der Flachheit der wandnormalen Geschwindigkeitsverteilung festzustellen. Diese Reynoldszahlabhängigkeiten sind auf die Interaktion der VLSMs mit dem wandnahen Strömungsbereich zurückzuführen.

Die folgende Analyse der axialen Energietransportgleichung des gefilterten Geschwindigkeitsfeldes legt offen, dass VLSMs Energie von der mittleren Strömung zugeführt bekommen, ähnlich den kleinskaligen Strukturen durch den turbulenten Produktionsmechanismus. Im Gegensatz zu den kleinskaligen Fluktuationen im wandnahen Zyklus, wird für VLSMs keine inverse Energiekaskade im statistischen Mittel detektiert. Skalen, welche kleiner sind als VLSMs sind jedoch verantwortlich für den physikalischen Transport von VLSM Energie in Richtung Wand und Hauptströmungsbereich. Zudem findet sich eine inverse Energiekaskade in Momentaufnahmen der Strömung in Regionen langsamer, sich von der Wand wegbewegender VLSMs. Dieses Verhalten unterscheidet sich grundsätzlich von dem kleinskaliger Fluktuationen, wo die inverse Energiekaskade mit schnellen, sich zur Wand hinbewegenden Strukturen zusammenhängt. In beiden Fällen ist der Energietransport schneller Strömungsbewegungen stärker, als jener der langsamen. Im Allgemeinen tritt die rückwärtige Energiekaskade vor allem in Regionen hoher turbulenter kinetischer Energie auf.

Zusammengefasst gewährt die Forschungsarbeit neue Einblicke in die Interaktion von VLSMs mit dem turbulenten Wandzyklus und trägt zum besseren Verständnis der turbulenten Rohrströmung bei.

Contents

List of Figures

List of Tables

Nomenclature

Roman Symbols

A_φ	constant in Townsend's prediction for the azimuthal Reynolds stress
A_z	constant in Townsend's prediction for the radial Reynolds stress
A_z	constant in Townsend's prediction for the streamwise Reynolds stress
B	log law constant
B_1	velocity-defect law constant
B_φ	coefficient in Townsend's prediction for the azimuthal Reynolds stress
B_z	coefficient in Townsend's prediction for the streamwise Reynolds stress
C_α	curvature term in the discretised momentum equation
D	pipe diameter
Δt	time step increment
E	instantaneous kinetic energy
e	instantaneous kinetic energy of the fluctuating velocity field
e^*	instantaneous kinetic energy of the filtered fluctuating velocity field
E_m	kinetic energy of the averaged flow field
$\mathbf{f}$	outer body forces
F	velocity flatness
f	probability density function

f_i	prefactors of the Leapfrog-Euler time integration scheme
h	channel half height
I	imaginary number
D_α	diffusive transport term in the discretised momentum equation
k	turbulent kinetic energy
K_α	convective transport term in the discretised momentum equation
Kn	Knudsen number
$\mathcal{L}$	characteristic length scale
L	pipe length
ℓ	length scale
ℓ_0	length scale of the largest eddies
ℓ_{DI}	demarcation length scale between dissipation and inertial subrange
ℓ_{EI}	demarcation length scale between energy-containing range and smaller eddies
$\mathbf{n}$	normal vector
N	number of time steps of one Leapfrog-Euler cycle
n_{proc}	number of computational cores
N_r	number of finite-volume cells in radial direction
N_z	number of finite-volume cells in axial direction
N_φ	number of finite-volume cells in azimuthal direction
$\mathcal{P}$	turbulent kinetic energy production
P	characteristic pressure scale
p	pressure
P_α	pressure term in the discretised momentum equation
$\mathbf{R}$	Reynolds stress tensor

$\mathbf{r}$	separation vector
R	pipe radius
r	radial coordinate
$R(s)$	auto-covariance
Re	Reynolds number
Re_b	bulk Reynolds number
Re_τ	friction Reynolds number
$R_{ij}(\mathbf{r}, \mathbf{x}, t)$	two-point velocity correlation
S	velocity skewness
S_{ij}	rate-of-strain tensor
$\mathcal{T}$	characteristic time scale
t	time
t_ν	viscous time scale
$\mathbf{u}^*$	auxiliary velocity field
$\mathcal{U}$	characteristic velocity scale
$\mathbf{u}$	velocity vector
U	sample space variable corresponding to u
u	streamwise velocity component
u_b	bulk velocity
u_η	Kolmogorov velocity scale
u_φ	azimuthal (spanwise) velocity component
u_i	i-component of the velocity
$u_{i,rms}$	root mean square of the i-velocity component
u_r	radial velocity component

u_τ friction velocity

u'_i i-component of fluctuating velocity

u_z axial (streamwise) velocity component

v vertical velocity component

w spanwise velocity component

x_i Cartesian coordinate of the i-direction

y wall distance

y^+ wall distance normalised by the viscous length scale

z axial coordinate

Greek Symbols

α inclination angle of velocity correlation iso-contours (chapter 4)

$\Delta\tau^+$ spatio-temporal averaging interval in viscous units

$\Delta\varphi$ azimuthal grid spacing

Δr_k radial grid spacing of finite-volume cells centred around r_k

ΔT temporal averaging interval

ΔT^+ averaging interval in viscous time units

Δt_{exp}^{crit} critical time step of the explicit Euler-Leapfrog scheme

Δt_{imp}^{crit} critical time step of the semi-implicit Euler-Leapfrog scheme

ΔT_b averaging interval in bulk time units

ΔV_k volume of a finite-volume cell centred around r_k

Δz axial grid spacing

δ boundary layer thickness

δ_ν viscous length scale

η Kolmogorov length scale

ε (pseudo-)dissipation of turbulent kinetic energy

ε_m dissipation rate of kinetic energy from the mean field

ϵ^2 mean square relative error

$\Phi_{ij}(\boldsymbol{\kappa}, t)$ velocity spectrum tensor

φ azimuthal coordinate

$\boldsymbol{\kappa}$ wave number vector

κ_i wave number w.r.t. the i-direction

κ von Kármán constant

λ_i wavelength w.r.t. the i-direction

λ_m mean free path

ν kinematic viscosity

ρ density

$\rho(s)$ auto-correlation coefficient

ρ_{ij} correlation coefficient between u_i and u_j

$\bar{\tau}$ integral time scale

τ_w wall shear stress

τ_{ij} viscous stress tensor

$\underline{\tau}$ viscous stress tensor in matrix notation

τ_{ij}^R residual stress tensor

τ_η Kolmogorov time scale

Mathematical Symbols

$\cdot$ scalar product

$\langle \phi \rangle$ average of quantity ϕ

$\mathrm{d}/\mathrm{d}x_i$ derivative in i-direction

$\partial/\partial x_i$	partial derivative in i-direction
$\otimes$	dyadic product
∇	Nabla operator
$\mathcal{O}$	order of magnitude
$\overline{\phi}$	spatially filtered quantity ϕ (chapter 5)
$\overline{\phi}$	surface/volume average of ϕ (chapter 2)
ϕ^+	quantity ϕ normalised in wall units
ϕ'	quantity ϕ-fluctuation $(=\phi - \langle\phi\rangle)$
$\tilde{\phi}$	residual of a spatially filtered quantity ϕ $(\tilde{\phi} = \phi - \overline{\phi})$

Acronyms / Abbreviations

CDF	cumulative distribution function
DNS	direct numerical simulation
FFT	fast Fourier transform
HPC	high-performance computing
JPDF	joint probability density function
LDA	laser Doppler anemometry
LES	large-eddy simulation
LSM	large-scale motions
MPI	message passing interface
PDF	probability density function
PIV	particle image velocimetry
POD	proper orthogonal decomposition
RANS	Reynolds-averaged Navier-Stokes
RMS	root mean square

SSM small-scale motions

TBL turbulent boundary layer

TKE turbulent kinetic energy

VLSM very-large-scale motions

Chapter 1

Introduction

We come across the phenomenon of turbulence in our everyday life in many different forms. As I am searching for the right words to begin this introduction, I gaze out of the window, where the clouds are passing by quite rapidly today. Being part of the atmospheric boundary layer, the advection of clouds is only one of the numerous examples of turbulent flow. Another one is the cup of coffee standing here in front of me[1]. When adding milk to the coffee, in order to mix the both of them, one usually does not wait until molecular diffusion does its job alone since its time scales are rather large. Inducing turbulence via stirring motion with a spoon, on the other hand, solves the problem. Turbulent transport creates a large interface between the coffee and the milk helping the diffusion process to be completed much faster than in the quiescent case. Besides these everyday-life examples, turbulence is of particular importance to numerous engineering applications, be it in the chemical processing industry, where turbulent flow supports the mixing between different chemicals — as it does for the coffee and the milk —, or in the field of aerodynamics, where turbulent flows around airplanes or cars cause an increase in drag, and thus, fuel consumption. To better understand viscous turbulence in the vicinity of solid walls, many experiments of the three canonical cases of wall turbulence — boundary layer, channel flow, and pipe flow — have been carried out. How essential the pipe flow experiment is in terms of turbulence research was pointed out by Richard Feynman, one of the greatest in physics, as follows:

> *The simplest form of the problem* [of turbulence] *is to take a pipe that is very long and push water through it at high speed. We ask: to push a given amount of water through that pipe, how much pressure is needed?*

[1] The contribution of that substance to the accomplishment of this thesis is highly acknowledged.

> *No one can analyse it from principles and the properties of water. If the water flows very slowly, or if we use a goo thick like honey, then we can do it nicely. You will find that in your textbook. What we really cannot do is deal with actual, wet water running through a pipe. That is the central problem which we ought to solve someday, and we have not.* — Feynman et al. (1963)

Today, we are still unable to solve the problem analytically, but with the aid of the computer resources available now, we can tackle it numerically. In this sense, I hope that the present work will serve as a tiny contribution to a better understanding of fluid turbulence in general. In the following section, the historical background of turbulent flow is further introduced with a special focus on turbulent pipe flow and direct numerical simulation (section 1.1), before the fundamentals of such flow are depicted in section 1.2. Finally, the introduction is closed with the goals and the outline of this thesis (section 1.3).

1.1 Background

The first and probably most famous study of turbulent pipe flow was performed by Reynolds (1883). He injected dye into the centreline of pipe water flow and discovered the onset of turbulence when increasing the flow rate above a certain level. Later, he identified the dependence of the transition to turbulence on a single non-dimensional parameter (Reynolds, 1894), which was then, in his honour, named Reynolds number

$$\mathrm{Re} = \frac{\mathcal{U}\mathcal{L}}{\nu}, \tag{1.1}$$

where $\mathcal{U}$ and $\mathcal{L}$ are the characteristic velocity and the characteristic length scale, respectively, and ν is the kinematic viscosity. The Reynolds number can be interpreted as the ratio between inertial and viscous forces in a fluid flow. With the increase of the Reynolds number, the dye in the pipe flow experiment leaves its shape of a straight line (laminar state) through a transition and becomes completely mixed (turbulent state, $\mathrm{Re} \gtrsim 2000$). The experimental set-up used by Reynolds (1883) as well as his observations of the laminar and turbulent state are depicted in figure 1.1. However, to this day, turbulence has not been clearly defined from a scientific point of view. Frisch (1995), for instance, stresses the ideas of *broken symmetries* and *restored symmetries* when describing the transition to turbulence. A more technical characterisation of turbulence is given by Mathieu and Scott (2000), who describe turbulence through

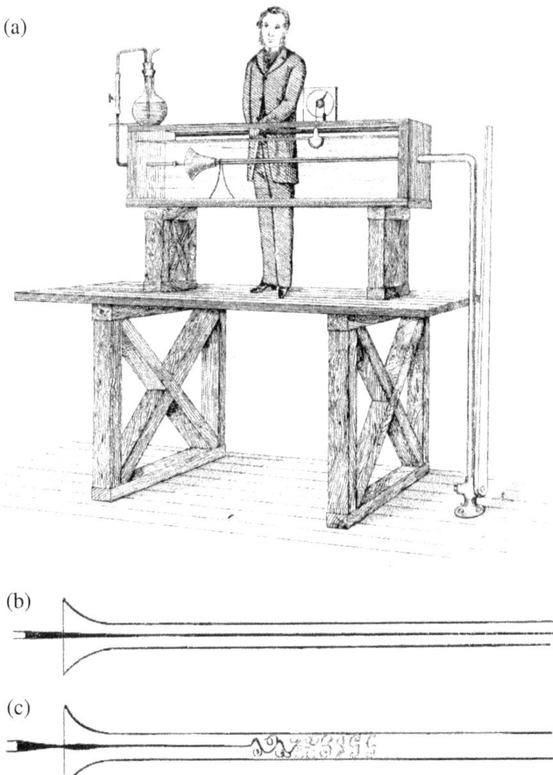

Figure 1.1: Experimental set-up used by Reynolds (1883) (a). Pipe with laminar (b) and turbulent (c) flow state. Figures taken from Reynolds (1883).

its features: *Turbulence is a random process.* Turbulent flows are dependent on space and time involving many degrees of freedom and are therefore unpredictable in detail. Turbulence statistics, however, appear to be reproducible. *Turbulence contains a wide range of scales.* While the largest scales of turbulent motion are determined by the flow geometry, the smallest scales depend on viscosity. As the Reynolds number increases, the range of turbulent scales expands. *Turbulence has small-scale random vorticity.* Space- and time-dependent, intensive, complex vortical motions can be found in turbulent flows with a high Reynolds number. *Turbulence arises at high Reynolds numbers.* The occurrence of turbulence is caused by the instability of a laminar flow. As the Reynolds number rises, the non-linear term of the governing flow equation becomes dominant with respect to the instability-damping viscous term. Hence, the flow tends towards the turbulent state. *Turbulence dissipates energy.* Small-scale turbulent motions cause large local velocity gradients. Consequently, the rate of kinetic energy dissipation into heat, which is proportional to the velocity gradients, is quite high in turbulent flows with respect to laminar flows. In statistically stationary turbulence, energy is continuously supplied to large-scale motions, from which it is transferred towards the smallest scales via the so-called turbulent energy cascade (Richardson, 1922), where it finally dissipates. *Turbulence is a continuum phenomenon.* The smallest scales in the turbulent flows considered in this thesis are some orders of magnitude larger than the molecular mean path. The governing equations describing the motion of fluids are the Navier-Stokes equations, see section 1.2.2. These non-linear equations feature strong sensitivity to small variations in the initial conditions and cannot be solved analytically without drastic simplifications. Due to the increase in the available computational power during the last decades, numerical methods have become important in the field of turbulence research. The most common approaches are referred to as Reynolds-averaged Navier-Stokes (RANS), large-eddy simulation (LES) and direct numerical simulation (DNS). The two first mentioned methods require turbulence modelling, since either only the time-averaged equations are solved, which introduces new unknown correlation terms (RANS), or only the larger turbulent scales are resolved (LES). Only the DNS fully resolves all spatial and temporal turbulent scales and is therefore the most accurate tool to study turbulence on all scales without having to make additional assumptions.

1.1.1 Direct Numerical Simulation

The term DNS was first used by Orszag (1970), who introduced this method to compute the energy spectrum of homogeneous isotropic incompressible turbulence.

DNS is not only the most accurate method of studying turbulence, but also the most expensive computationally, since the computational costs increase as Re^3 (Pope, 2000). Consequently, the first DNSs were restricted to flows with low Reynolds number. In addition to homogeneous turbulent flows with or without imposed mean velocity gradients (Bardina et al., 1985; Lee et al., 1990; Orszag and Patterson, 1972; Rogers and Moin, 1987), inhomogeneous flows such as turbulent boundary layer (TBL) as well as turbulent pipe and plane channel flows have also been subject to DNSs. Here, steep velocity gradients and small-scale motions in the vicinity of the wall require a higher resolution towards the solid boundary. The numerical studies of turbulent plane channel flow performed by Kim et al. (1987), of TBL flow performed by Spalart (1988), and of turbulent pipe flow performed by Eggels et al. (1994) were pioneering DNSs in wall-bounded turbulence and have given us deep insights into the so-called turbulent near-wall cycle (see section 1.2.5). The investigation of the near-wall cycle by means of DNS was later the focus of many studies, i.e. Jeong et al. (1997); Jiménez et al. (2005); Jiménez and Moin (1991); Jiménez and Pinelli (1999).

With the improvement of computational resources, simulations with higher Reynolds numbers became feasible and have been carried out in the last decades. High Reynolds numbers exhibit a broad range of turbulent scales, and it was discovered that not only the small scales related to the near-wall cycle contain energy, but also the large scales located in the logarithmic region. Channel flow DNSs by Abe et al. (2004, 2001); del Álamo and Jiménez (2003); del Álamo et al. (2004); Hoyas and Jiménez (2006); Jiménez et al. (2004) dealt with these large-scale motions and their interaction with the wall — described in more detail in section 1.2.5 — by extending the Reynolds number regime up to $\mathrm{Re}_\tau = 2003$. Please note that this is the friction Reynolds number, which is defined later in equation (1.46). Recently, in order to reach a clear distinction between the small-scale motions (SSMs) near the wall and the large- and very-large-scale motions (LSMs,VLSMs) in the outer layer and to observe a distinguished logarithmic region in plane channel flow, DNSs have been carried out for $\mathrm{Re}_\tau = 5200$ by Lee and Moser (2015) and for $\mathrm{Re}_\tau = 8000$ by Yamamoto and Tsuji (2018). Besides the above-mentioned, the scope of DNSs done on wall-bounded turbulence comprises open channel flows (Handler et al., 1999, 1993; Nagaosa, 1999; Nagaosa and Handler, 2003; Wang and Richter, 2019), open and closed duct flows (Gavrilakis, 1992; Sakai, 2019; Uhlmann et al., 2007), plane Couette flows (Avsarkisov et al., 2014; Lee and Moser, 2018; Pirozzoli et al., 2014), and the turbulent Ekman layer in the planetary atmosphere (Ansorge and Mellado, 2014), amongst other cases. A brief overview of the DNSs of turbulent pipe flow can be found in the next section together with experimental investigations of this flow type.

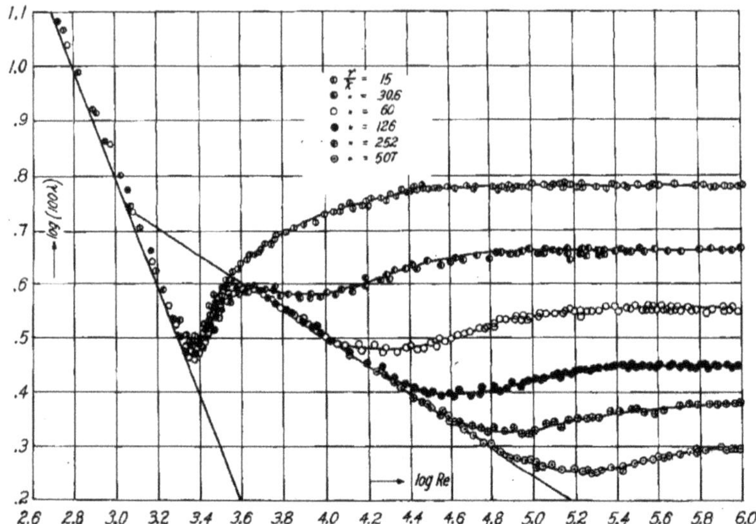

Figure 1.2: Relation between log(100λ) and log(Re) with λ being the friction factor, k/r the relative roughness and Re the Reynolds number. Figure adapted from NACA technical memorandum 1292, a translation of Nikuradse (1933).

1.1.2 Turbulent Pipe Flow

Pipe flows occur in natural as well as in technical systems. The flow of blood through human vessels is an example of a mostly laminar[2] non-Newtonian pipe flow. Most technical applications, on the other hand, feature turbulent flows of Newtonian fluids. Consequently, the turbulent flow of a Newtonian fluid through a pipe has been the subject of various experimental and numerical studies. After Reynolds (1883), pioneering investigations focussing on friction laws in turbulent pipe flow were carried out by Blasius (1913) and later by Nikuradse (1932, 1933), whose relations between the friction factor, the Reynolds number, and the pipe roughness, as shown in figure 1.2, have become basic knowledge for today's engineering students. Moreover, Nikuradse (1932) was the first to fit the constants of the *logarithmic law of the wall* (or *log law*) by von Kármán (1930) — introduced in section 1.2.5 — into experimental data. While

[2]Indeed, the flow of blood through the vessels of a healthy human being is laminar. Diseases as arteriosclerosis, however, cause a local partial blockage of the vessel, which leads to a turbulent blood flow.

Nikuradse (1933) used an artificial uniform roughness by means of a sand coating in the pipe interior, Colebrook and White (1937) repeated the measurement with non-uniform roughness and derived a friction law for the transition between the smooth and the fully rough regime. Finally, the findings of the studies of Blasius (1913), Nikuradse (1932, 1933), and Colebrook and White (1937) have been visualised in the so-called Rouse diagram (after Rouse (1943)[3]), which also found its way into today's engineering textbooks (see e.g. Schlichting and Gersten (2017)). Henceforth, I only consider the case of a turbulent flow through a smooth pipe, as described in section 1.2.1.

The first detailed measurement of turbulence statistics involving the mean velocity profile, second-, third-, and fourth-order moments of the velocity distribution as well as velocity spectra were performed by Laufer (1954) with the aid of hot-wire anemometry. As the latter study exhibited some inconsistencies regarding the measurement of the dissipation rate, Lawn (1971) later reported a more accurate determination of the dissipation rate in turbulent pipe flow. Patel and Head (1969) measured the log law constants experimentally for both turbulent plane channel and pipe flow. They found the constants to vary between those two types of flow and attributed this to the different geometry. Aiming at the scaling of statistics in turbulent pipe flows, Perry and Abell (1975) carried out detailed measurements over a range of Reynolds numbers ($80000 \leq \mathrm{Re} \leq 260000$) and compared them with previous works. They described an inner and an outer scaling region for turbulence intensities, which is discussed in more detail in section 1.2.5. The scaling of turbulence statistics is closely related to the underlying turbulent coherent structures. Sabot and Comte-Bellot (1976) investigated coherent structures in the core region of turbulent pipe flow by means of a quadrant analysis of the Reynolds shear stress. While the structure of the Reynolds shear stress was found to be similar to the structure in turbulent boundary-layer flow in the vicinity of the wall, this was not the case for the core region. Here, the structures on one side of the pipe axis interacted with the structures from the other side, which produced different results with respect to the turbulent boundary-layer flow. Perry et al. (1986) developed a well-recognised model for wall turbulence including the scaling of turbulence spectra, which they underpinned with pipe flow measurements.

In the 1990s, new measurement techniques, such as laser Doppler anemometry (LDA) and particle image velocimetry (PIV) came into play. Westerweel et al. (1993) and Durst et al. (1995b) worked on highly resolved near-wall measurements of turbulent

[3]The diagram is (in a slightly different form) also referred to as the Moody chart, after Moody (1944). However, although Moodys adaption uses more practicable coordinates (λ and Re instead of $1/\sqrt{\lambda}$ and $\mathrm{Re}\sqrt{\lambda}$), it contains the same information as the original work. Consequently, I prefer to introduce the original Rouse diagram here.

pipe flow with a low Reynolds number by means of PIV and LDA, respectively. These experimental studies were in agreement with plane channel flow DNS (Kim et al., 1987) for statistical moments of the velocity distribution up to fourth order, in general. However, discrepancies were detected for the flatness factor of the radial velocity near the wall, which was not as steep as in the DNS data.

It was not until 1993 that the first DNSs of turbulent pipe flow were performed by Eggels et al. (1993) and Unger (1994). A good comparison between their DNS data of $\mathrm{Re}_\tau = 180$ and both the turbulent pipe flow experiments and the plane channel flow DNS can be found in Eggels et al. (1994). Their work confirmed earlier findings by Patel and Head (1969), who presented different log laws for channel and pipe flow. Most of the turbulence statistics up to fourth order, on the other hand, agreed well between the DNSs of both types of flow. Only the wall-normal skewness factor was found to differ. Eggels et al. (1994) connected this phenomenon to a different "impingement" mechanism close to the curved pipe wall. The high level of wall-normal flatness very close to the wall found in numerical simulations (both channel and pipe) was again absent in the experimental data. Furthermore, the digital PIV study by Westerweel et al. (1996), who experimented on the same Reynolds number case with a new measurement technique, was perfectly in line with earlier experimental studies, and thus, did not recover the high wall-normal flatness level obtained from numerical observations.

One of the biggest controversies in wall-bounded turbulence in the latest history was the question whether the assumptions on which the log law is based are correct, and whether the log law is universal, meaning it stands independent of the Reynolds number. Barenblatt (1993); Barenblatt and Prostokishin (1993) questioned the validity of the log law and proposed a power law of the following form instead

$$u^+ = C \cdot (y^+)^\alpha, \tag{1.2}$$

with the coefficients C and α. To evaluate the different theories of the mean velocity scaling for very high Reynolds numbers, turbulent pipe flow experiments have been carried out in the *superpipe* facility in Princeton, USA, over a wide range of Reynolds numbers. While Barenblatt et al. (1997) argued that the *superpipe* data underpin the theory of a power law, Zagarola et al. (1997) suggested that the data rather support a universal log law for high Reynolds numbers. Figure 1.3 displays the mean velocity profiles measured by Zagarola and Smits (1997) over a range of Reynolds numbers and the log law they fitted. Additionally, Zagarola and Smits (1997) fitted the power law to their data. Both fits led to an accurate representation of the mean velocity

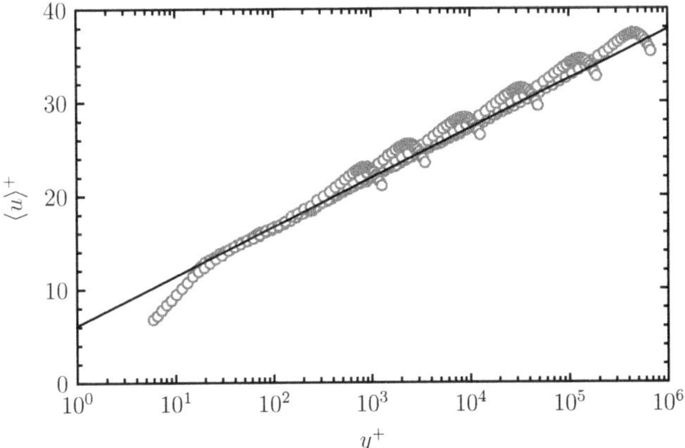

Figure 1.3: Mean velocity profiles of fully-developed turbulent pipe flow. ○, Experimental data of Zagarola and Smits (1997), $\mathrm{Re} \approx 3.2 \times 10^4$, 9.9×10^4, 4.1×10^5, 1.8×10^6, 7.7×10^6, and 2.99×10^7. Solid line, log law $u^+ = 1/\kappa y^+ + B$ with $\kappa = 0.436$ and $B = 6.13$.

profiles. However, while the constant α in the power law depends on the Reynolds number, the constants of the log law are independent of the Reynolds number, which means that the log law is universal. As a consequence, Pope (2000) highlights the practical importance of the log law as compared to the power law. Despite the type of the mean velocity scaling law and its universality, there is an ongoing debate about its range of validity. While the overlap region was initially believed to start around $y^+ \approx 30$, this value has been shifted more and more away from the wall. Accordingly, Zagarola and Smits (1998) suggested that the logarithmic region starts at $y^+ \approx 600$ for high-Reynolds-number pipe flow. Besides, Zagarola and Smits (1998) proposed a power law region located before the log law region in terms of wall distance. Later, McKeon et al. (2004a,b, 2005) repeated the experiment using more accurate measurement techniques and confirmed the findings of Zagarola and Smits (1998) reporting only slightly different log law coefficients. By comparing the superpipe data with data obtained from boundary layer and channel flow experiments, Nagib and Chauhan (2008) discovered the von Kármán coefficient to depend on the flow type, and thus, to be non-universal. Almost a decade later, Luchini (2017) showed that the dependency on the flow type is caused by the effect of the pressure gradient on the mean velocity profile. Taking this effect into account by means of a higher-order perturbation, he derived a novel universal formulation of the log law.

Another very influential investigation in terms of turbulent pipe flow is the study of Kim and Adrian (1999), who measured long regions of streamwise velocity fluctuations they called VLSMs. These turbulent coherent structures contribute to the turbulent kinetic energy and interact with the small viscous scales in the vicinity of the wall. As a consequence, experimental (Bailey and Smits, 2010; Guala et al., 2006; Hellström et al., 2011; Hellström and Smits, 2014) as well as numerical investigations (Ahn et al., 2013, 2015; Lee et al., 2015; Lee and Sung, 2013; Wu et al., 2012) aimed at a structural characterisation of VLSMs in turbulent pipe flow, which is further described in section 1.2.5. While the first turbulent pipe flow DNSs by Eggels et al. (1993) and Unger (1994) were restricted to Reynolds numbers as low as $\mathrm{Re}_\tau = 180$, the rapid increase of computational resources within the last decades has led to a successive increase in Reynolds numbers in the studies of turbulent pipe flow and turbulent plane channel flow. Wagner et al. (2001) examined the low-Reynolds-number effects in turbulent pipe flow up to $\mathrm{Re}_\tau = 320$, before Wu and Moin (2008) carried out pipe flow DNS up to $\mathrm{Re}_\tau = 1142$. Whereas the latter studies involved second-order schemes for the spatial and temporal discretisation, Boersma (2011, 2013) computed velocity statistics up to $\mathrm{Re}_\tau = 1842$ in fully-developed turbulent pipe flow using a pseudo-spectral method, and El Khoury et al. (2013) used a spectral-element method for pipe flow DNS up to $\mathrm{Re}_\tau = 1000$. Recently, Ahn et al. (2017, 2013, 2015), Lee and Sung (2013), and Lee et al. (2015) have carried out turbulent pipe flow DNSs using a second-order, finite-difference scheme and have extended the regime of Reynolds numbers accessible by DNSs up to $\mathrm{Re}_\tau = 3008$.

The debate whether the mean velocity profile follows a log law or a power law, as well as the investigation of the interaction of VLSMs with the near-wall turbulence, calls for highly-resolved, high-Reynolds-number experiments. Since the above-described experiments were performed in the Princeton superpipe facility, several additional facilities for turbulent pipe flow experiments have been set up — such as the Cottbus large pipe test facility in Germany (CoLaPipe, König et al., 2014; Öngüner, 2018), the high-Reynolds-number flow facility in Japan (Hi-Reff, Furuichi et al., 2015), and the centre for international cooperation in long pipe experiments in Italy (CICLoPE, Fiorini, 2017; Talamelli et al., 2009). Thanks to its large diameter ($D = 0.9\,m$), the CICLoPE facility is capable of measuring the smallest turbulent scales even for high-Reynolds-number flows, see figure 1.4. Recently, researches have been conducted in this facility studies regarding the scaling of turbulent fluctuations (Örlü et al., 2017; Willert et al.,

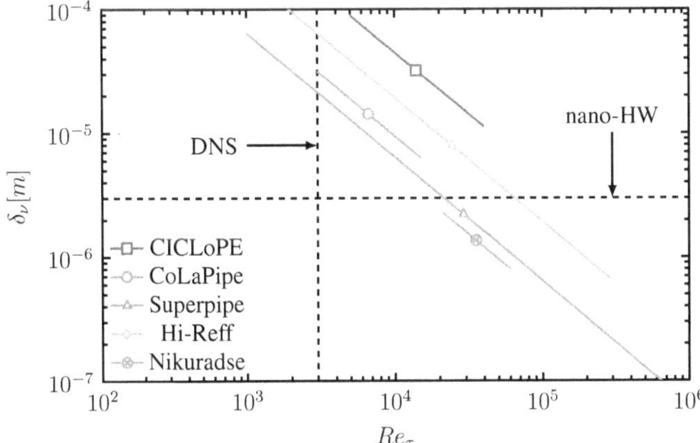

Figure 1.4: Viscous length scale δ_ν versus friction Reynolds number Re_τ for different experimental facilities. The vertical dashed line shows the currently highest Reynolds number pipe flow DNS by Ahn et al. (2015). The horizontal dashed line indicates the currently smallest hot wire probes introduced by Bailey et al. (2010). Figure adapted from Fiorini (2017).

2017), the characterisation of VLSMs in high-Reynolds-number flow (Discetti et al., 2019), the simultaneous measurement of full velocity and vorticity vectors (Zimmerman et al., 2019), and the skin friction measurement (Baidya et al., 2019). Despite the improvement in the measurement resolution in the pipe flow experiments with high Reynolds number, the small-scale structure of the near-wall turbulence — scaling with the viscous length scale —, as well as its interaction with larger turbulent scales originating further away from the wall, is still difficult to obtain from experiments. Hence, in order to complement the experimental investigations, fully-resolved DNSs of turbulent pipe flow are required to answer the open research questions stated in section 1.3. However, before answering those questions, the fundamentals of turbulent pipe flow will be introduced in section 1.2.

1.2 Fundamentals

In the upcoming section, the problem of turbulent pipe flow is conceptually introduced by means of the geometry of the flow (section 1.2.1) and the flow-governing equations (section 1.2.2). In addition, the description of turbulence by means of statistical

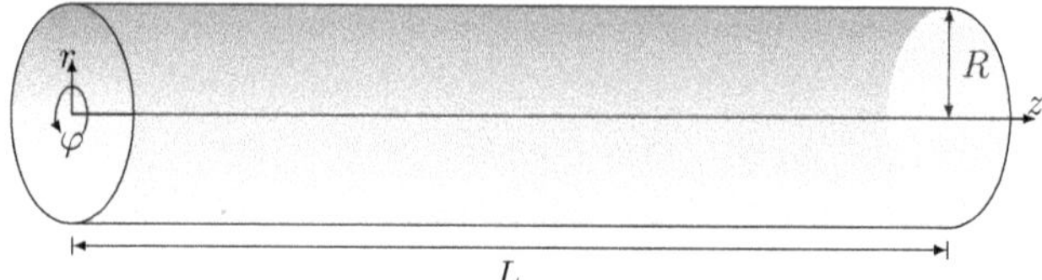

Figure 1.5: Pipe geometry with a cylindrical coordinate system, where z is the axial, φ the azimuthal, and r the radial coordinate. L is the pipe length and R the pipe radius.

quantities is presented (section 1.2.3), followed by a deeper exploration of the features of wall-bounded turbulence (section 1.2.5). Finally, the phenomenology of the kinetic energy transport between coherent structures of different sizes in turbulent flows is recounted in section 1.2.6.

1.2.1 Flow Geometry

The flow geometry is described by a smooth annular pipe with the length L and the pipe radius R, as shown in figure 1.5. The problem is illustrated in a cylindrical coordinate system, where z is the axial, φ the azimuthal, and r the radial coordinate. The flow geometry is determined by periodic boundaries in the axial direction ($z = 0$ and $z = L$), since the flow is assumed to be statistically homogeneous in this direction. Moreover, a smooth no-slip boundary condition, as well as an impermeability boundary condition, is applied at the pipe wall ($r = R$).

1.2.2 Equations of Fluid Motion

Assumptions

For the present flow cases, the following assumptions are made:

1. Using the Knudsen number $\mathrm{Kn} = \lambda_m/\mathcal{L} \ll 1$ with the mean free path λ_m as a microscopic and the characteristic length $\mathcal{L}$ as a macroscopic scale, the flow of interest can be treated as a continuum.

2. Furthermore, the flow of interest is incompressible, which means that the density remains constant along the path of a fluid particle.

3. The fluid is considered a Newtonian fluid, whose strain rate is linear proportional to the viscous stress, which is true for most of the fluids in technical applications.

The Newtonian ansatz is

$$\tau_{ij} = \rho\nu\left(\frac{\partial u_i}{\partial x_j} + \frac{\partial u_j}{\partial x_i}\right), \tag{1.3}$$

where τ_{ij} is the viscous stress tensor.

Governing Equations

Considering the conservation of mass and momentum, the governing equations, also referred to as Navier-Stokes equations[4], read

$$\frac{\partial \rho}{\partial t} + \nabla \cdot (\rho \mathbf{u}) = 0, \tag{1.4a}$$

$$\frac{\partial \rho \mathbf{u}}{\partial t} + \nabla(\rho(\mathbf{u} \otimes \mathbf{u})) = -\nabla p + \nabla \underline{\tau} + \rho \mathbf{f}, \tag{1.4b}$$

with the velocity $\mathbf{u} = (u, v, w)$, the density ρ, the pressure p, the viscous stress tensor $\underline{\tau}$, and outer body forces $\mathbf{f}$ (e.g. gravitation). Mathematical operators are defined in appendix A. Applying the aforementioned assumptions, the transport equations for mass and momentum for an incompressible, continuous flow of a Newtonian fluid,

$$\nabla \cdot \mathbf{u} = 0, \tag{1.5a}$$

$$\frac{\partial \mathbf{u}}{\partial t} + (\mathbf{u} \cdot \nabla)\mathbf{u} = -\frac{1}{\rho}\nabla p + \nu \nabla^2 \mathbf{u} + \mathbf{f}, \tag{1.5b}$$

are obtained. For a constant density (i.e. no thermal effects), the gravity can — according to Kundu and Cohen (2008) — be absorbed into a modified pressure variable, which yields

$$\nabla \cdot \mathbf{u} = 0, \tag{1.6a}$$

$$\frac{\partial \mathbf{u}}{\partial t} + (\mathbf{u} \cdot \nabla)\mathbf{u} = -\frac{1}{\rho}\nabla p + \nu \nabla^2 \mathbf{u}. \tag{1.6b}$$

In order to non-dimensionalise the equations, a characteristic length scale $\mathcal{L}$ and a characteristic velocity scale $\mathcal{U}$ have to be chosen. The reference values for time and pressure are then $\mathcal{T} = \mathcal{L}/\mathcal{U}$ and $P = \rho\mathcal{U}^2$. By multiplying equation (1.6a) with $\mathcal{L}/\mathcal{U}$

[4]Claude-Louis Navier derived the original equation in 1823 (Navier, 1823). Later, in 1845, Sir George Gabriel Stokes included internal fluid friction (Darrigol, 2009).

and equation (1.6b) with $\mathcal{L}/\mathcal{U}^2$, we obtain the non-dimensional equations

$$\nabla \cdot \tilde{\mathbf{u}} = 0, \tag{1.7a}$$

$$\frac{\partial \tilde{\mathbf{u}}}{\partial \tilde{t}} + (\tilde{\mathbf{u}} \cdot \nabla)\tilde{\mathbf{u}} = -\nabla \tilde{p} + \frac{1}{\mathrm{Re}} \nabla^2 \tilde{\mathbf{u}}, \tag{1.7b}$$

with the Reynolds number Re as defined in (1.1) and the non-dimensional variables

$$\tilde{\mathbf{u}} = \frac{\mathbf{u}}{\mathcal{U}}, \; \tilde{p} = \frac{p}{P}, \; \tilde{\mathbf{x}} = \frac{\mathbf{x}}{\mathcal{L}}, \; \tilde{t} = \frac{t}{\mathcal{T}}.$$

Hereinafter, the tilde for dimensionless parameters is dropped, and transport equations are considered to be dimensionless. Choosing the pipe radius R as characteristic length scale and the friction velocity u_τ — introduced in equation (1.43) — as characteristic time scale, the Navier-Stokes equations for an incompressible flow of a Newtonian fluid are

$$\nabla \cdot \mathbf{u} = 0, \tag{1.8a}$$

$$\frac{\partial \mathbf{u}}{\partial t} + \mathbf{u} \cdot \nabla \mathbf{u} + \nabla p = \frac{1}{\mathrm{Re}_\tau} \nabla^2 \mathbf{u}, \tag{1.8b}$$

where $\mathrm{Re}_\tau = u_\tau R/\nu$ is the friction Reynolds number. Using indicial notation and the Einstein summation convention, the Cartesian coordinate representatives of the above equations read as follows:

$$\frac{\partial u_i}{\partial x_i} = 0, \tag{1.9a}$$

$$\frac{\partial u_j}{\partial t} + u_i \frac{\partial u_j}{\partial x_i} = -\frac{\partial p}{\partial x_j} + \frac{1}{\mathrm{Re}_\tau} \frac{\partial^2 u_j}{\partial x_i^2}, \; j = 1,2,3. \tag{1.9b}$$

In cylindrical coordinates (z,φ,r) with u_z being the axial, u_φ the azimuthal, and u_r the radial velocity, equations (1.8a) and (1.8b) equal to

$$\frac{\partial u_z}{\partial z} + \frac{1}{r}\frac{\partial u_\varphi}{\partial \varphi} + \frac{1}{r}\frac{\partial}{\partial r}(u_r r) = 0, \tag{1.10a}$$

$$\frac{\partial u_z}{\partial t} + u_z \frac{\partial u_z}{\partial z} + \frac{1}{r} u_\varphi \frac{\partial u_z}{\partial \varphi} + u_r \frac{\partial u_z}{\partial r} + \frac{\partial p}{\partial z} = \frac{1}{\mathrm{Re}_\tau} \nabla^2 u_z, \tag{1.10b}$$

$$\begin{aligned} \frac{\partial u_\varphi}{\partial t} &+ u_z \frac{\partial u_\varphi}{\partial z} + \frac{1}{r}\left(u_\varphi \frac{\partial u_\varphi}{\partial \varphi} + u_r u_\varphi \right) + u_r \frac{\partial u_\varphi}{\partial r} + \frac{1}{r}\frac{\partial p}{\partial \varphi} \\ &= \frac{1}{\mathrm{Re}_\tau}\left[\nabla^2 u_\varphi - \frac{1}{r^2}\left(u_\varphi - 2\frac{\partial u_r}{\partial \varphi} \right) \right], \end{aligned} \tag{1.10c}$$

$$\begin{aligned} \frac{\partial u_r}{\partial t} &+ u_z \frac{\partial u_r}{\partial z} + \frac{1}{r}\left(u_\varphi \frac{\partial u_r}{\partial \varphi} - u_\varphi^2 \right) + u_r \frac{\partial u_r}{\partial r} + \frac{\partial p}{\partial r} \\ &= \frac{1}{\mathrm{Re}_\tau}\left[\nabla^2 u_r - \frac{1}{r^2}\left(u_r + 2\frac{\partial u_\varphi}{\partial \varphi} \right) \right], \end{aligned} \tag{1.10d}$$

with

$$\nabla^2 \phi = \frac{\partial^2 \phi}{\partial z^2} + \frac{1}{r}\frac{\partial}{\partial r}\left(r \frac{\partial \phi}{\partial r} \right) + \frac{1}{r^2}\frac{\partial^2 \phi}{\partial \varphi^2}, \tag{1.11}$$

see e.g. Batchelor (1967).

1.2.3 Statistical Description of Turbulence

Due to the chaotic behaviour of turbulent flow, turbulent quantities are treated statistically. The probabilistic description of turbulence dates back to Kolmogorov (1933) and Taylor (1935, 1938). Later, Monin and Yaglom (1971, 1975) published a comprehensive work on the statistical description of turbulence. Moreover, the English language standard reference on the statistical description of turbulence was published by Tennekes and Lumley (1972). A more recent introduction to the topic can be found in Pope (2000). Following the notation of Pope, the statistical quantities relevant to this work are defined in this section. First, we define the random variable u as a representative of a turbulent quantity and the sample space variable U, which characterises the range of values that u could possibly represent. Considering the event B that u is smaller than one specific value U_b, viz. $B = \{u < U_b\}$, leads to an expression for the probability that the event B is true:

$$p = P(B) = P\{u < U_b\}, \tag{1.12}$$

where $0 \leq p \leq 1$. The probability of any event in the sample space is determined by the cumulative distribution function (CDF) defined as

$$F(U) = P\{u < U\}. \tag{1.13}$$

The derivative of the CDF is denoted as the probability density function (PDF)

$$f(U) = \frac{dF(U)}{dU}. \tag{1.14}$$

Angular brackets denote the mean of the random variable,

$$\langle u \rangle = \int_{-\infty}^{\infty} U f(U) dU, \tag{1.15}$$

or in general the mean of a function $Q(u)$,

$$\langle Q(u) \rangle = \int_{-\infty}^{\infty} Q(U) f(U) dU. \tag{1.16}$$

Furthermore, the fluctuation of u with respect to its mean is defined as

$$u' = u - \langle u \rangle, \tag{1.17}$$

and the variance of u is its mean square fluctuation,

$$var(u) = \langle (u')^2 \rangle = \int_{-\infty}^{\infty} (U - \langle u \rangle)^2 f(U) dU. \tag{1.18}$$

The square root of the variance is the standard deviation σ_u and also denoted as the root mean square (RMS) of u:

$$u_{rms} = \sigma_u = \sqrt{var(u)} = \sqrt{\langle u'u' \rangle}. \tag{1.19}$$

The nth-order statistical moment of u reads

$$\langle (u')^n \rangle = \int_{-\infty}^{\infty} (U - \langle u \rangle)^n f(U) dU. \tag{1.20}$$

The third- and fourth-order standardised central moments

$$S(u) = \langle u'^3 \rangle / u_{rms}^3, \tag{1.21}$$

$$F(u) = \langle u'^4 \rangle / u_{rms}^4, \tag{1.22}$$

are denoted as skewness and flatness (or kurtosis). Problems involving more than one random variable — that are not necessarily stochastically independent — require joint probability functions (JPDF). For two random variables u_1, u_2 the JPDF reads

$$f_{12}(U_1, U_2) = \frac{\partial^2}{\partial U_1 \partial U_2} F_{12}(U_1, U_2), \tag{1.23}$$

with $F_{12}(U_1, U_2) = P\{u_1 < U_1, u_2 < U_2\}$. The mean of a function Q of the random variables is defined as

$$\langle Q(u_1, u_2)\rangle = \int_{-\infty}^{\infty}\int_{-\infty}^{\infty} Q(U_1, U_2)_{12} f(U_1, U_2) dU_1 dU_2. \tag{1.24}$$

Means or variances can be derived from equation (1.24). The covariance of u_1 and u_2,

$$cov(u_1, u_2) = \langle u_1' u_2'\rangle = \int_{-\infty}^{\infty}\int_{-\infty}^{\infty} (U_1 - \langle u_1\rangle)(U_2 - \langle u_2\rangle) f(U_1, U_2) dU_1 dU_2, \tag{1.25}$$

or their correlation coefficient, respectively,

$$\rho_{12} = \frac{\langle u_1' u_2'\rangle}{\sqrt{\langle u_1' u_1'\rangle\langle u_2' u_2'\rangle}}, \tag{1.26}$$

is a measure for the statistical dependency between the two random variables. In the case of time-dependent random variables — the so-called random processes — the CDF and PDF are expanded by the temporal dimension. Here, the N-time joint CDF F_N or PDF f_N carry information about the temporal correlation of the random variables. For statistically stationary random processes, F_N and f_N are time-invariant, i.e.

$$f_N(U_1, t_1 + T, U_2, t_2 + T, ...) = f_N(U_1, t_1, U_2, t_2, ...), \tag{1.27}$$

and the auto-covariance can be defined as

$$R(s) = \langle u'(t) u'(t+s)\rangle \tag{1.28}$$

or normalised as the auto-correlation coefficient

$$\rho(s) = \frac{R(s)}{\langle u'(t)^2\rangle}, \tag{1.29}$$

which is expected to decay to zero for large time separations. The integral time scale

$$\bar{\tau} = \int_0^\infty \rho(s)ds \tag{1.30}$$

measures how fast the process decorrelates in time. Besides their dependence on time, turbulent flows are described by three-dimensional vector fields. Therefore, statistical quantities depend on their spatial position. Consequently, the one-point, one-time PDF of a random variable u reads

$$f(U,\mathbf{x},t) = \frac{\partial F(U,\mathbf{x},t)}{\partial U}, \tag{1.31}$$

with $F(U,\mathbf{x},t) = P\{u(\mathbf{x},t) < U\}$ being the one-time, one-point CDF. In statistically homogeneous turbulence, the N-point PDF is invariant to a shift in space,

$$f_N(U,\mathbf{x}_1+\mathbf{X},\mathbf{x}_2+\mathbf{X},...) = f_N(U,\mathbf{x}_1,\mathbf{x}_2,...). \tag{1.32}$$

Since the PDF is usually unknown, mean quantities are obtained by averaging over a number of samples. For a sufficiently high number of samples, the average approaches the value of the probabilistic average[5]. In fully-developed turbulent pipe flow, advantage of the statistical stationarity and homogeneity can be taken by averaging over time as well as over both homogeneous directions. Accordingly, angular brackets indicate an average as follows:

$$\langle u\rangle(r) = \frac{1}{L}\frac{1}{2\pi r}\frac{1}{\Delta T}\int_{t=t_0}^{t_0+\Delta T}\int_{z=0}^{L}\int_{\varphi=0}^{2\pi} u(z,\varphi,r,t)r d\varphi dz dt. \tag{1.33}$$

The two-point, one-time auto-covariance — or the two-point correlation — of a random field reads

$$R_{ij}(\mathbf{r},\mathbf{x},t) = \langle u_i'(\mathbf{x},t)u_j'(\mathbf{x}+\mathbf{r},t)\rangle, \tag{1.34}$$

where $\mathbf{r}$ is the separation vector between the two points. For statistical stationarity, the dependence on t, and for statistical homogeneity, the dependence on $\mathbf{x}$ vanishes. In spatially homogeneous directions, turbulence quantities can be expressed in Fourier space representation. The Fourier transform of the two-point velocity covariance R_{ij}

[5]The number of samples required for certain statistical quantities to converge is determined in section 4.2.1.

reads

$$\Phi_{ij}(\boldsymbol{\kappa},t) = \frac{1}{(2\pi)^3}\int_{-\infty}^{\infty} e^{-I\boldsymbol{\kappa}\cdot\mathbf{r}} R_{ij}(\mathbf{r},t)d\mathbf{r}, \tag{1.35}$$

and is referred to as the velocity spectrum tensor with $\boldsymbol{\kappa} = (\kappa_1, \kappa_2, \kappa_3)$ being the wave number vector, whose components $\kappa_i = 2\pi/\lambda_i$ are related to the wavelengths of the Fourier mode λ_i, and with $I = \sqrt{-1}$. The inverse transform of velocity spectrum tensor yields

$$R_{ij}(\mathbf{r},t) = \int_{-\infty}^{\infty} e^{I\boldsymbol{\kappa}\cdot\mathbf{r}} \Phi_{ij}(\boldsymbol{\kappa},t)d\boldsymbol{\kappa}. \tag{1.36}$$

By setting $\mathbf{r}$=0 and using statistical stationarity and homogeneity, an expression for one-dimensional wave number spectra with respect to either stream- or spanwise direction is obtained as follows:

$$R_{ii}(0) = \langle u_i' u_i' \rangle = \int_{\kappa_x=0}^{\infty} \Phi_{ii}(\kappa_x)d\kappa_x = \int_{\kappa_z=0}^{\infty} \Phi_{ii}(\kappa_z)d\kappa_z. \tag{1.37}$$

Analogously, expressions for the so-called pre-multiplied, one-dimensional wavelength spectra can be derived through

$$R_{ii}(0) = \langle u_i' u_i' \rangle = \int_{\log\lambda_x=0}^{\infty} \kappa_x \Phi_{ii}(\kappa_x)d\log\lambda_x = \int_{\log\lambda_z=0}^{\infty} \kappa_z \Phi_{ii}(\kappa_z)d\log\lambda_z, \tag{1.38}$$

with $\lambda_x = 2\pi/\kappa_x$ being the stream- and $\lambda_z = 2\pi/\kappa_z$ the spanwise wavelength.

1.2.4 The Scales of Turbulent Motions

As described in chapter 1, turbulence involves different scales of motions. Richardson (1922) introduced the turbulent *energy cascade*, where turbulence is considered to be a composition of eddies of different sizes ℓ. In this framework, the largest eddies, with a length scale ℓ_0 comparable to the characteristic length scale of the flow $\mathcal{L}$ and with a velocity scale u_0, obtain energy via the production mechanism. Subsequently, energy is transferred to increasingly smaller scales by an inviscid mechanism until it finally dissipates from the smallest eddies due to molecular viscosity. Later, Kolmogorov (1941) picked up the idea of the Richardson cascade, postulating his well-known hypotheses. First, **Kolmogorov's hypothesis of local isotropy** reads:

At sufficiently high Reynolds number, the small-scale turbulent motions ($\ell \ll \ell_0$) are statistically isotropic.

While large scales are influenced by the flow geometry and boundary conditions, this information disappears when energy is distributed towards smaller scales. Consequently, the statistics of small-scale motions are universal for high-Reynolds-number flows. Kolmogorov (1941) defined the length scale ℓ_{EI} as a demarcation between the so-called *energy-containing range* ($\ell > \ell_{EI}$), where eddies contain large portions of energy and are highly anisotropic, and the *universal equilibrium range*, where the dominant processes are the transfer of energy to smaller scales and viscous dissipation. Hence, **Kolmogorov's first similarity hypothesis** states:

At sufficiently high Reynolds number, the statistics of the small-scale motions ($\ell < \ell_{EI}$) have a universal form that is uniquely determined by kinematic viscosity ν and dissipation rate ε.

The characteristic scales obtained from the parameters ν and ε,

$$\eta = \left(\frac{\nu^3}{\varepsilon}\right)^{(1/4)}, \tag{1.39}$$

$$u_\eta = (\varepsilon\nu)^{(1/4)}, \tag{1.40}$$

$$\tau_\eta = \left(\frac{\nu}{\varepsilon}\right)^{(1/2)}, \tag{1.41}$$

which are the characteristic scales of the smallest turbulent motions, are denoted as Kolmogorov scales. Assuming $\varepsilon \approx u_0^3/\ell_0$, the ratio between the largest and smallest scales of turbulent motion can be derived as

$$\frac{\ell_0}{\eta} \approx \mathrm{Re}^{3/4}, \tag{1.42}$$

showing that the range of scales increases with the Reynolds number. For a very-high-Reynolds-number flow, there is a scale ℓ that is very small compared with ℓ_0, though being still very large when compared with η. Accordingly, **Kolmogorov's second similarity hypothesis** maintains:

At sufficiently high Reynolds number, the statistics of the motions of scale ℓ in the range $\eta \ll \ell \ll \ell_0$ have a universal form that is uniquely determined by ε and independent of kinematic viscosity ν.

In accordance with the *second similarity hypothesis*, the universal equilibrium range can be divided into two subranges, namely the inertial subrange ($\ell_{EI} > \ell > \ell_{DI}$), where motions are determined by *inertial effects*, and the *dissipation range* ($\ell < \ell_{DI}$), where viscous effects — and thus, dissipation — play a role. Figure 1.6 portrays a

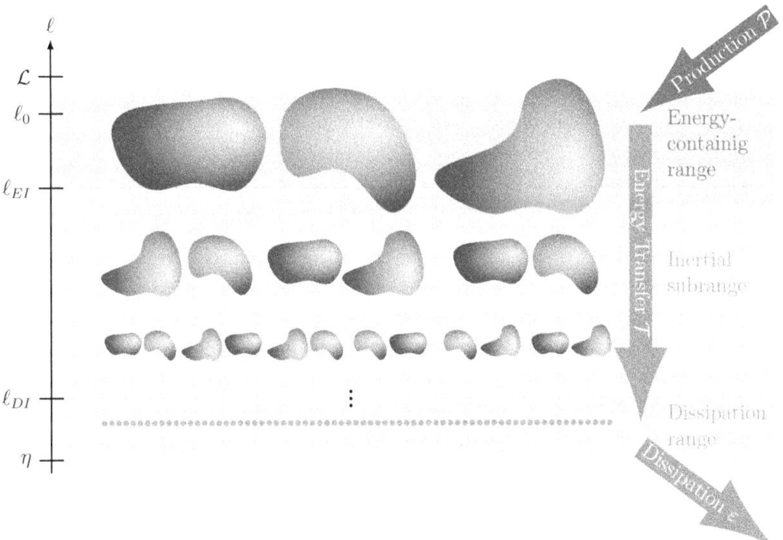

Figure 1.6: Sketch of the turbulent energy cascade after Richardson (1922) including length scales introduced by Kolmogorov (1941).

scheme of the *energy cascade* and the different scales involved. One consequence of the second similarity hypothesis is a universal $-5/3$ power law in the inertial subrange of the velocity power spectrum ($\phi_{ii} \sim \kappa^{-5/3}$). While the small scales show a universal behaviour in turbulent flows, the large scales strongly depend on the flow geometry, as discussed above. In most technical applications, turbulence interacts with solid walls.

1.2.5 Wall-bounded Turbulence

Besides plane channel and flat-plate boundary layer flow, turbulent pipe flow is one of the canonical examples of wall-bounded turbulence. Since turbulence dissipates a large amount of the energy needed to move a fluid through a pipe or a vehicle through air in the vicinity of a wall, the investigation of wall-bounded turbulence is of utmost importance to the field of aerodynamics. Again, for a deeper introduction to the topic, refer to Pope (2000). A fully-developed turbulent pipe flow is characterised by two homogeneous directions parallel to the wall — the axial and the azimuthal direction — and by the inhomogeneous wall-normal direction. As indicated in equation (1.33), turbulence one-point statistics depend only on the radial coordinate (r) in such flows. Further, according to Prandtl (1925), wall-bounded turbulence can be divided into an inner region close to the solid wall and an outer region further away from the solid wall. In the inner region, the viscosity plays an important role, and thus, characteristic scales of the near-wall region are based on the viscosity and are referred to as *viscous scales*. The characteristic velocity scale of the inner region is the *friction velocity*

$$u_\tau = \sqrt{\frac{\tau_w}{\rho}}, \tag{1.43}$$

with $\tau_w = \rho\nu|\partial\langle u\rangle/\partial r|_{r=R}$ representing the wall shear stress. Moreover, the *viscous length scale* is defined as

$$\delta_\nu = \frac{\nu}{u_\tau}, \tag{1.44}$$

and the *viscous time scale* as

$$t_\nu = \frac{\nu}{u_\tau^2}. \tag{1.45}$$

The characteristic dimensionless quantity describing wall-bounded turbulent flows is the friction Reynolds number

$$\mathrm{Re}_\tau = \frac{u_\tau R}{\nu} = \frac{R}{\delta_\nu}, \tag{1.46}$$

which is expressed as the ratio between the outer-flow length scale R and the inner viscous length scale δ_ν. Normalisation in viscous scales — also referred to as "in wall units" is hereinafter denoted with the superscript "+". Accordingly, the distance from the wall,

$$y = R - r, \tag{1.47}$$

in wall units reads

$$y^+ = (R-r)^+ = \frac{R-r}{\delta_\nu} = \mathrm{Re}_\tau - \frac{r}{\delta_\nu}, \tag{1.48}$$

and the mean axial velocity in wall units is defined by

$$u^+ = \frac{\langle u_z\rangle}{u_\tau}. \tag{1.49}$$

From dimensional analysis, Prandtl derived that in the vicinity of the wall, the mean velocity depends only on the viscous length scale, i.e.

$$u^+ = f_w(y^+), \tag{1.50}$$

which is also known as the *law of the wall.* By applying a Taylor-series expansion to the no-slip boundary condition $u^+ = 0$ at the solid wall $y^+ = 0$, the linear relation

$$u^+ = y^+ + \mathcal{O}(y^{+2}) \tag{1.51}$$

can be derived. The region where deviations of the mean velocity from $u^+ = y^+$ are negligibly small ($y^+ < 5$) is designated as *viscous sublayer.* For large Re_τ, the outer part of the inner layer ($r/R \ll 1$) corresponds to large y^+ ($y^+ \gg 1$). In this region, the logarithmic law of the wall or *log law* according to von Kármán (1930),

$$u^+ = \frac{1}{\kappa}\log y^+ + B, \tag{1.52}$$

with the constant B and the von Kármán constant κ, is valid. As described in section 1.1.2, the questions whether the *log law* holds true and where it is valid are controversial issues up to today. Nevertheless, Pope (2000) estimates the range of validity of the *log law* from $y^+ > 30$ to $y/R < 0.3$. The region above the viscous sublayer and below the logarithmic layer ($5 < y^+ < 30$) is the *buffer layer.* In the outer layer, where direct viscosity effects are negligibly small, the relevant length scale is the pipe radius R[6]. For the outer-flow region, von Kármán (1930) developed the *velocity-defect law,*

$$\frac{u_0 - \langle u_z \rangle}{u_\tau} = F_D\left(\frac{y}{R}\right), \tag{1.53}$$

which describes the difference between the mean velocity $\langle u_z \rangle$ and its centreline value $u_0 = \langle u_z \rangle(r=0)$. For high-Reynolds-number flows, an *overlap region* between the inner layer ($y/R < 0.1$) and the outer layer ($y^+ > 50$) exists, where both the *log law* and the *velocity defect law* are valid. Following Millikan (1938), the *velocity-defect law* in this layer can be established as

$$\frac{u_0 - \langle u_z \rangle}{u_\tau} = -\frac{1}{\kappa}\log\left(\frac{y}{R}\right) + B_1, \text{ for } \frac{y}{R} \ll 1, \tag{1.54}$$

with B_1 being a flow-dependant constant. An overview of the different layers and regions in wall-bounded turbulence adapted from Pope (2000) is given in figure 1.7. Although defined via the mean velocity profile, characteristics of the different layers and regions in wall-bounded turbulence are also determined by fluctuating quantities.

[6]For plane channel turbulence, the characteristic length scale of the outer-flow region would be the channel half-height h, and for flat-plate boundary layer flows, the boundary layer thickness δ. Hence, the mean velocity in the outer-flow region — unlike in the inner region — highly depends on the flow geometry.

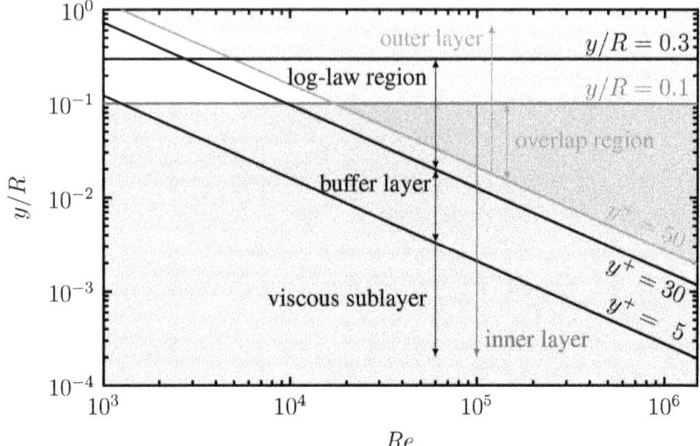

Figure 1.7: Regions and layers in turbulent pipe flow as functions of the Reynolds number. Figure adapted from (Pope, 2000).

In the upcoming sections, the characteristics of wall-bounded turbulence are discussed by means of the turbulent kinetic energy, the small-scale dynamics of the near-wall cycle, the attached eddy model, the implications of LSMs and VLSMs, as well as *velocity spikes*.

The Balance of Turbulent Kinetic Energy

Integrating the Reynolds decomposition (1.17) into the definition of the fluids kinetic energy, i.e.

$$E = \frac{1}{2} u_i u_i, \tag{1.55}$$

and taking the average (1.33) results in the decomposition of the mean kinetic energy into two parts,

$$\langle E \rangle = \underbrace{\frac{1}{2}\langle u_i \rangle \langle u_i \rangle}_{E_m} + \underbrace{\frac{1}{2}\langle u_i' u_i' \rangle}_{k}, \tag{1.56}$$

where E_m is the energy contained in the average flow field and k is the kinetic energy of the turbulent fluctuations (TKE). Averaging the Navier-Stokes equations (1.9) brings in the *Reynolds equations*,

$$\frac{\partial \langle u_i \rangle}{\partial x_i} = 0, \tag{1.57a}$$

$$\frac{\partial \langle u_j \rangle}{\partial t} + \frac{\partial \langle u_i \rangle \langle u_j \rangle}{\partial x_i} + \frac{\partial \langle u_i' u_j' \rangle}{\partial x_i} = \frac{1}{\mathrm{Re}_\tau} \frac{\partial^2 \langle u_j \rangle}{\partial x_i \partial x_i} - \frac{\partial \langle p \rangle}{\partial x_j}, \; j = 1,2,3, \tag{1.57b}$$

here given in Cartesian coordinates. By subtracting equation (1.57b) from equation (1.9b) and multiplying the result with the fluctuating velocity u_j', one can deduce a transport equation for the turbulent kinetic energy:

$$\left(\frac{\partial k}{\partial t} + \langle u_j \rangle \frac{\partial k}{\partial x_j} \right) + \frac{1}{2} \frac{\partial \langle u_j' u_i' u_i' \rangle}{\partial x_j} - \frac{1}{\mathrm{Re}_\tau} \frac{\partial^2 k}{\partial x_j \partial x_j} + \left\langle u_i' \frac{\partial p'}{\partial x_i} \right\rangle = \mathcal{P} - \varepsilon. \tag{1.58}$$

Multiplying equation (1.57b) with $\langle u_j \rangle$, on the other hand, delivers a transport equation for the kinetic energy of the mean velocity

$$\left(\frac{\partial E_m}{\partial t} + \langle u_j \rangle \frac{\partial E_m}{\partial x_j} \right) + \frac{\partial}{\partial x_j} \left[\langle u_i \rangle \langle u_i' u_j' \rangle - \frac{2}{\mathrm{Re}_\tau} \langle u_i \rangle \langle S_{ij} \rangle + \langle u_i \rangle \langle p \rangle \right] = -\mathcal{P} - \varepsilon_m, \tag{1.59}$$

where

$$S_{ij} = \frac{1}{2} \left(\frac{\partial u_i}{\partial x_j} + \frac{\partial u_j}{\partial x_i} \right) \tag{1.60}$$

is the rate-of-strain tensor, and

$$\mathcal{P} = -\langle u_i' u_j' \rangle \frac{\partial \langle u_i \rangle}{\partial x_j} \tag{1.61}$$

is the production term, which is generally positive. It represents the work performed by the gradients of the mean field against the Reynolds stresses and appears in both transport equations (1.58,1.59) with opposite signs. It is, thus, an exchange term between the kinetic energy of the mean and the kinetic energy of the fluctuating field. Since it is generally positive, it removes energy from the mean flow transferring it to the fluctuations. The dissipation rate of kinetic energy from the mean flow field reads as follows:

$$\varepsilon_m = \frac{2}{\mathrm{Re}_\tau} \langle S_{ij} \rangle \langle S_{ij} \rangle, \tag{1.62}$$

while the dissipation rate due to the fluctuating field,

$$\varepsilon = \frac{1}{\mathrm{Re}_\tau} \left\langle \frac{\partial u_i'}{\partial x_j} \frac{\partial u_i'}{\partial x_j} \right\rangle, \tag{1.63}$$

is termed as turbulent dissipation[7] and acts as a sink in the transport equation of turbulent kinetic energy. For fully-developed turbulent pipe flow, the TKE budget equation in cylindrical coordinates,

$$\frac{1}{2r}\frac{\mathrm{d}(r\langle u_r' u_i' u_i'\rangle)}{\mathrm{d}r} - \frac{1}{\mathrm{Re}_\tau}\left[\frac{1}{r}\frac{\mathrm{d}}{\mathrm{d}r}\left(r\frac{\mathrm{d}k}{\mathrm{d}r}\right)\right] + \frac{1}{r}\frac{\mathrm{d}(r\langle u_r' p'\rangle)}{\mathrm{d}r} = \mathcal{P} - \varepsilon, \tag{1.64}$$

with

$$\mathcal{P} = -\langle u_r' u_z'\rangle \frac{\partial\langle u_z\rangle}{\partial x_r} \tag{1.65}$$

being the TKE production term and with

$$\begin{aligned}\varepsilon = -\frac{1}{\mathrm{Re}_\tau}\Bigg[&\left\langle\left(\frac{\partial u_r'}{\partial r}\right)^2\right\rangle + \left\langle\frac{1}{r^2}\left(\frac{\partial u_r'}{\partial\varphi} - u_\varphi'\right)^2\right\rangle + \left\langle\left(\frac{\partial u_r'}{\partial z}\right)^2\right\rangle \\ &+ \left\langle\left(\frac{\partial u_\varphi'}{\partial r}\right)^2\right\rangle + \left\langle\frac{1}{r^2}\left(\frac{\partial u_\varphi'}{\partial\varphi} + u_r'\right)^2\right\rangle + \left\langle\left(\frac{\partial u_\varphi'}{\partial z}\right)^2\right\rangle \\ &+ \left\langle\left(\frac{\partial u_z'}{\partial r}\right)^2\right\rangle + \left\langle\frac{1}{r^2}\left(\frac{\partial u_z'}{\partial\varphi}\right)^2\right\rangle + \left\langle\left(\frac{\partial u_z'}{\partial z}\right)^2\right\rangle\Bigg]\end{aligned} \tag{1.66}$$

being the TKE dissipation term, is derived from the cylindrical coordinate representation of the Navier-Stokes equations (1.10) in appendix B. The transport terms of the left-hand side of equation (1.64) are denoted as turbulent diffusion (TD), viscous diffusion (VD), and pressure diffusion (PD). Figure 1.8(a) illustrates the TKE budget terms for fully-developed turbulent pipe flow. While the peak of TKE production is located in the buffer layer at $y^+ \approx 12$, dissipation is highest at the solid wall. The transport terms are responsible for the distribution of TKE in physical space. The viscous diffusion term acts closer to the wall than the turbulent diffusion term does. Both terms mainly transport TKE away from the location of the TKE production peak — where it is induced to the fluctuating field — towards the wall, where it is dissipated via the dissipation term. The production term is a correlation between the viscous stress and the Reynolds shear stress, both of which contribute to the total shear stress. The profiles of the shear stress contribution can be taken from figure 1.8(b). The Reynolds shear stress is negligibly small within the viscous sublayer, whereas viscous effects do not play a role further away from the wall. Consequently, the TKE production peak appears in the buffer layer, where a large mean velocity gradient exists together with a large Reynolds shear stress. The evolution of the Reynolds shear stress

[7] To be precise, this is the pseudo-dissipation. For another decomposition between viscous-diffusion and dissipation the reader is referred to Pope (2000), page 132.

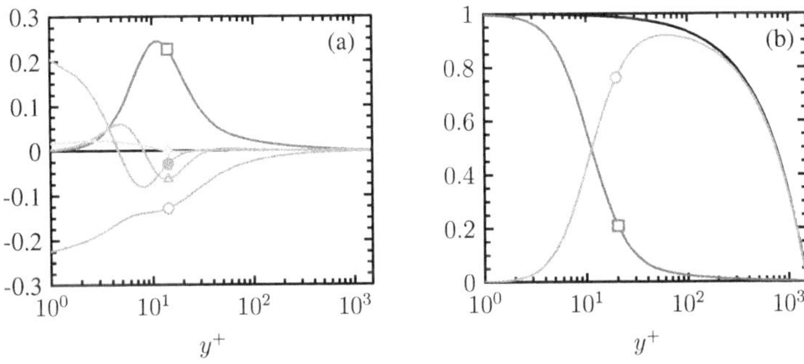

Figure 1.8: (a) Turbulent kinetic energy budget terms normalised in wall units: -□-, production ($\mathcal{P}$); -○-, dissipation (ε); -●-, viscous diffusion (VD); -·-, pressure diffusion (PD); -△-, turbulent diffusion (TD). (b) Decomposition of the total shear stress (—) into viscous stress $\mathrm{d}\langle u_z\rangle^+/\mathrm{d}y^+$ (-□-) and Reynolds shear stress $\langle u_z' u_r'\rangle^+$ (-○-) in wall units. Pipe flow DNS data, $\mathrm{Re}_\tau = 1500$.

profile as well as the normal Reynolds stress profiles are often discussed in terms of the underlying coherent structures.

The Near-wall Cycle

The structure of wall-bounded turbulence is determined by different coherent motions that appear in the vicinity of the wall. The so-called *sweeps* and *ejections* contribute significantly to the Reynolds shear stress (Brodkey et al., 1974; Kim et al., 1987; Wallace et al., 1972; Willmarth and Lu, 1972). While the term *sweeps* refers to events of high-speed fluid moving toward the wall ($u_z' > 0$, $u_r' > 0$), events of low-speed fluid moving away from the wall are usually denoted as *ejections* ($u_z' < 0$, $u_r' < 0$). The JPDF of the axial and radial velocity components, depicted in figure 1.9 for a wall distance of $y^+ = 15$, shows *sweeps* and *ejections* being more probable than other events. Based on their quadrant[8] in the two-dimensional sample space, *sweeps* and *ejections* are also denoted as Q4 or Q2 events, respectively. Like the Reynolds shear stress, the structure of the normal Reynolds stress component is determined by the underlying turbulent coherent structures. The near-wall peak of TKE (at $y^+ \approx 12$) is caused by strong elongated streamwise velocity fluctuations, also referred to as *velocity-* or *wall-layer streaks*, which were first observed by Kline et al. (1967). Amongst others,

[8]The quadrant analysis is an important tool for the investigation of wall-bounded turbulence. For deeper insight the reader is referred to the review paper of Wallace (2016) and the references therein.

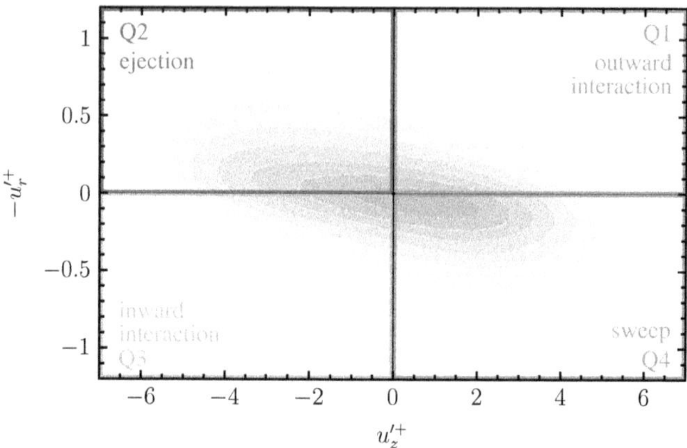

Figure 1.9: Joint probability density function (JPDF) $f(u_z'^+, u_r'^+)$ at a wall distance of $y^+ = 15$. Iso-surfaces ranging from 0.02 (light) to 0.14 (dark) with an increment of 0.02. Pipe flow DNS data, $\mathrm{Re}_\tau = 1500$.

Robinson (1991) characterises these motions as alternating *high-* ($u_z' > 0$) and *low-speed streaks* ($u_z' < 0$) with a spacing of 100 viscous units (between either two low- or two high-speed streaks) and a length of the order $\mathcal{O}(1000\delta_\nu)$. *Sweeps* and *ejections* are closely associated with the so-called *quasi-streamwise vortices*, which, according to Blackwelder and Eckelmann (1979), play an important role in the formation of streaks. Further away from the wall, in the logarithmic layer, the appearance of *sweeps* and *ejections* is often related to the dynamics of arch-shaped or *hairpin vortices* introduced by Theodorsen (1952) and first noticed by Head and Bandyopadhyay (1981). A conceptual model encompassing the last mentioned coherent structures and their relation, sketched by Robinson (1991), is depicted in figure 1.10. Swearingen and Blackwelder (1987) describe the growth and breakdown of streamwise vortices as a part of a regeneration cycle of near-wall turbulence (or *near-wall cycle*). By artificially removing outer-flow fluctuations from a numerical simulation, Jiménez and Pinelli (1999) showed that the near-wall cycle is independent of the outer flow. The cycle consists of the generation of streamwise vortices via the instability of streaks and the formation of velocity streaks from the advection of the mean profile by streamwise vortices. Motions related to the near-wall cycle, which scale in viscous units, such as velocity streaks or quasi-streamwise vortices, are in the following subsumed under the term *small-scale motions* (SSM). Another type of coherent motions connected to the

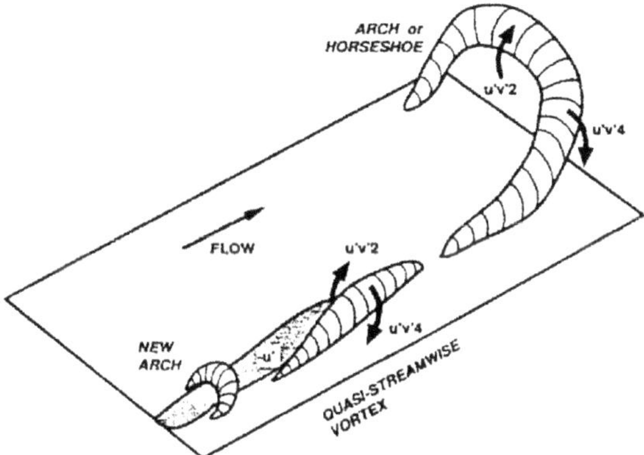

Figure 1.10: Conceptual model of near-wall coherent structures in low-Reynolds-number TBLs. $u'v'2$, ejections; $u'v'4$, sweeps. Figure adapted from Robinson (1991).

near-wall cycle are very large, local, wall-normal velocity fluctuations, which are rare both in space and time. Xu et al. (1996) refer to them as *velocity spikes* and found them to be responsible for the high wall-normal kurtosis level in the vicinity of the wall. Velocity spikes and their influence on turbulence statistics will be explained in depth at the end of this section.

The Attached Eddy Model

To describe the flow characteristics of the logarithmic layer, Townsend (1976) introduced the *attached eddy model.* Within this framework, the turbulence structure is modelled as a superposition of wall-attached eddies which grow with their distance from the wall. These anisotropic motions carry a substantial fraction of Reynolds shear stress and are therefore regarded as "active". The so-called *wall-detached eddies*, on the contrary, which are more isotropic and carry only little Reynolds shear stress, are referred to as "inactive" motions. Assuming a random detached eddy distribution, Townsend (1976) derived scaling laws for the Reynolds stress components in the high-Reynolds-number

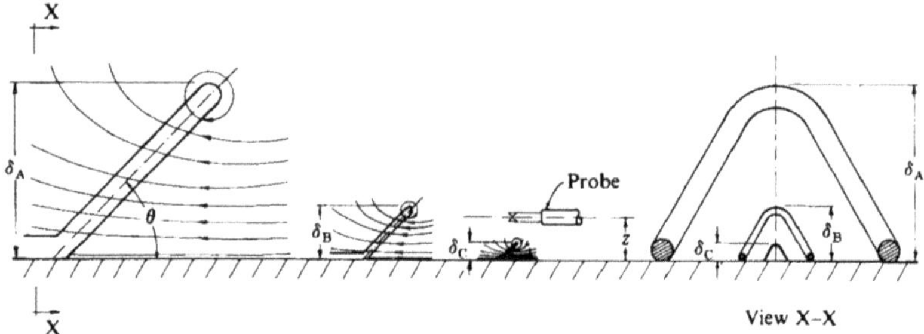

Figure 1.11: Sketch of three attached eddies of different scales. Figure adapted from Perry et al. (1986)

limit reading

$$\begin{aligned} \langle u'_z u'_z \rangle^+ &= A_z - B_z \log(y/R), &(1.67)\\ \langle u'_\varphi u'_\varphi \rangle^+ &= A_\varphi - B_\varphi \log(y/R), &(1.68)\\ \langle u'_r u'_r \rangle^+ &= A_r, &(1.69)\\ \langle u'_z u'_r \rangle^+ &= 1. &(1.70)\end{aligned}$$

The attached eddy model was expanded by Perry and fellow researchers (Perry and Chong, 1982; Perry et al., 1986; Perry and Marušic, 1995), who described attached eddies as hairpin-like geometries depicted as in figure 1.11. Based on their model, Perry et al. (1986) derived scaling laws for velocity spectra in the logarithmic layer, which they verified via experimental results. For streamwise velocity fluctuations, Perry et al. (1986) proposed three different spectral scaling regions. First, eddies of scale $l = \mathcal{O}(R)$ contribute only to low wave numbers, which leads to an *outer-flow* scaling of the low-wave number part of the spectrum,

$$\frac{\phi_{zz}(\kappa_z R)}{u_\tau^2} = g_1(\kappa_z R). \qquad (1.71)$$

Then, eddies that scale with the distance from the wall $l = \mathcal{O}(y)$ would contribute to an intermediate range of wave numbers, i.e.

$$\frac{\phi_{zz}(\kappa_z y)}{u_\tau^2} = g_2(\kappa_z y). \qquad (1.72)$$

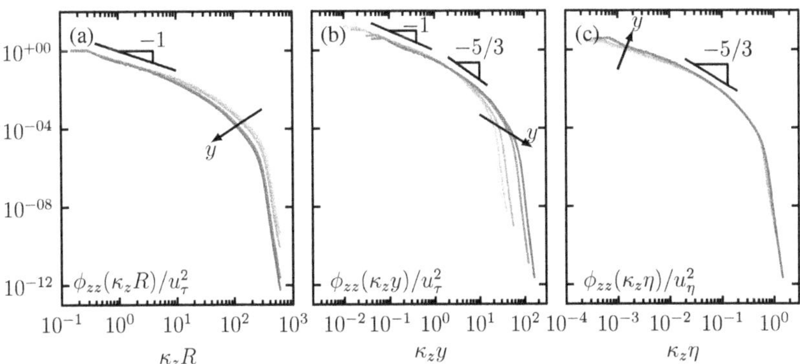

Figure 1.12: Scaling of the streamwise velocity spectrum in the logarithmic layer according to Perry et al. (1986). (a) Outer scaling region for low wave numbers $\phi_{zz}(\kappa_z R)$. (b) Inner scaling region for intermediate wave numbers $\phi_{zz}(\kappa_z y)$. (c) Kolmogorov scaling for large wave numbers $\phi_{zz}(\kappa_z \eta)$. Spectra in the logarithmic layers at wall distances of $y^+ = 79, 99, 141, 299, 401$ with colours from light blue ($y^+ = 79$) to dark blue ($y^+ = 401$). Pipe flow DNS data, $\mathrm{Re}_\tau = 1500$.

Last, reaffirming Kolmogorov (1941), in the section of the spectrum with very high wave numbers, where motions are locally isotropic and viscosity-dependent, the spectrum scales as

$$\frac{\phi_{zz}(\kappa_z \eta)}{u_\eta^2} = g_3(\kappa_z \eta). \tag{1.73}$$

Additionally, Perry et al. (1986) anticipated two regions of overlap, where either expressions (1.71) and (1.72) or expressions (1.72) and (1.73) are simultaneously valid. While the latter overlap region corresponds to the aforementioned inertial subrange, where the spectrum follows the Kolmogorov -5/3 law $\phi_{zz} \sim \kappa_z^{-5/3}$, Perry et al. (1986) derived a power law scaling of $\phi_{zz} \sim \kappa_z^{-1}$ for the former overlap region. Figure 1.12 shows the scaling of streamwise velocity spectra in the logarithmic region of turbulent pipe flow of $\mathrm{Re}_\tau = 1500$. Moreover, Perry et al. (1986) introduced "pre-multiplied" velocity spectra, where the energy contribution of a wave number range corresponds to the area under the spectrum. Consequently, the most energy-containing motions can be deduced from the peak location of the pre-multiplied spectrum. In the pre-multiplied velocity spectrum, the κ_z^{-1} range forms a plateau, whose large and small wave number shoulder collapses in inner and outer-flow scaling coordinates, respectively. This small wave number shoulder or outer spectral peak is due to the appearance of LSMs and VLSMs in wall-bounded turbulent flows.

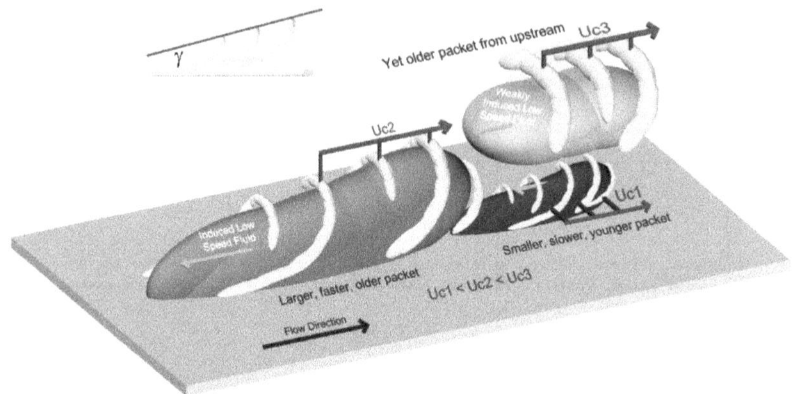

Figure 1.13: Conceptual model of hairpin-vortex organisation in turbulent boundary layers. Figure adapted from Adrian (2007).

Large-Scale Motions

The former type of coherent structures served also as a base for the conceptual scenario of *hairpin vortex packets* introduced by Adrian et al. (2000) and depicted in figure 1.13(b). Adrian et al. (2000) described the formation of turbulent bulges in TBLs, which are also referred to as LSMs, as a consequence of the streamwise alignment of hairpin vortices resulting in hairpin vortex packets. These packets grow from the wall and induce approximately uniform momentum zones. Their propagation velocity is determined by the velocity of their environment, so that packets that reach further away from the wall are faster than the ones close to the wall, which is also represented in figure 1.13. As a consequence, the model features hierarchies of smaller, slower packets nested in larger, faster packets. The ultimate dimension of the packets, which is in the order of the size of LSMs, is finally reached in the wake region. LSMs have first been measured in TBLs, where a mean streamwise extent of approximately $(2-3)\delta$ and a spanwise extent of approximately $(1-1.5)\delta$ were reported for a boundary layer thickness of δ (Brown and Thomas, 1977; Kovasznay et al., 1970; Murlis et al., 1982).

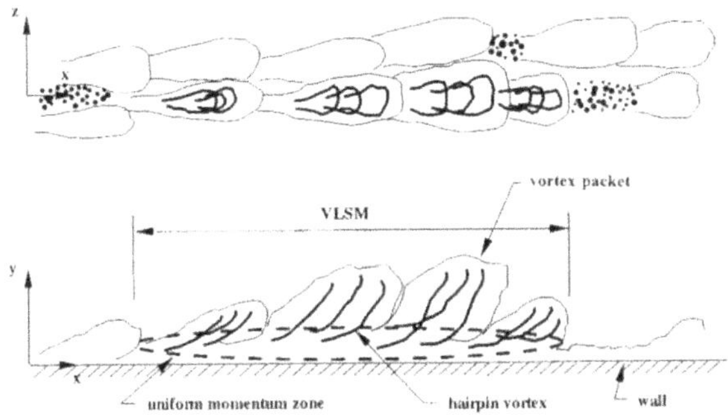

Figure 1.14: Conceptual model of VLSM and the process of vortex packet alignment proposed by Kim and Adrian (1999). Figure adapted from Kim and Adrian (1999).

Very-Large-Scale Motions

By analysing pre-multiplied energy spectra, Kim and Adrian (1999) recognised regions of streamwise velocity fluctuations extending up to 14 pipe radii in the logarithmic region of turbulent pipe flow, which they referred to as VLSMs. Kim and Adrian (1999) identified VLSMs by decomposing velocity spectra with a κ^{-1} range into two modes, which scale with inner and outer units, respectively. Here, the outer mode wavelength corresponds to VLSMs. Hereinafter, VLSMs are either identified by the location of the large wavelength shoulder of the κ^{-1} range or by an outer spectral peak in pre-multiplied energy spectra if there is one. Moreover, Kim and Adrian (1999) suggested that VLSMs are created by a streamwise coherent alignment of several hairpin vortex packets, as presented in figure 1.14. When analysing plane channel flow DNS data of Kim et al. (1987) and of Moser et al. (1999), Jiménez (1998) detected very long streamwise wavelength contributions to the streamwise velocity spectrum, which were constrained by the numerical box. Later, del Álamo and Jiménez (2003) carried out plane channel DNS in large numerical boxes — $12\pi h$ and $8\pi h$ for $\mathrm{Re}_\tau = 180$ and $\mathrm{Re}_\tau = 550$, respectively — where they found large fractions of energy in structures longer than $\lambda_z = 5h$, which they referred to as very large anisotropic scales. Guala et al. (2006) measured both LSMs ($2R - 3R$) and VLSMs (more than $8R - 16R$) in experimental turbulent pipe flow of $3815 \leq \mathrm{Re}_\tau \leq 7959$. Besides, Guala et al. (2006) found VLSMs to carry about half of the total turbulent kinetic energy and more than half of the Reynolds shear stress, which means that VLSMs are to be considered as *active* motions. Monty et al. (2009, 2007) compared the large-scale features among the different cases of wall-bounded turbulence — turbulent pipe, channel, and boundary layer flow — for a similar Reynolds number $\mathrm{Re}_\tau \approx 3000$. While LSMs were found to be of approximately the same length in all three cases — $\lambda_z \approx 3\delta$ with δ either the boundary layer thickness, the channel half height or the pipe radius —, VLSMs were found to be much longer in internal flows — $14 \leq \lambda_z/\delta \leq 20$ — than in boundary layer flow — $\lambda_z/\delta \approx 6$. Following Hutchins and Marusic (2007a), the large streamwise structures in TBL flow are referred to as *superstructures* to distinguish them from even larger VLSMs in turbulent pipe and channel flow. Hellström et al. (2011) proved via proper orthogonal decomposition (POD) and instantaneous flow field visualisations that VLSMs already exist in pipe flows at Reynolds numbers as low as $\mathrm{Re}_b = 12500$. Hutchins and Marusic (2007b), Smits et al. (2011), and Rosenberg et al. (2013), however, showed that VLSMs become increasingly energetic as Re goes up, although they exist at all Reynolds numbers. The interaction between VLSMs or superstructures and small-scale motions in the near-wall region has been the focus of many studies. Hutchins and Marusic (2007b) and Mathis et al. (2009a, 2011, 2009b) coined the concept of *amplitude modulation*, which describes the modulation of the small-scale signal near the wall by outer-layer LSMs and VLSMs. Furthermore, Abe et al. (2004) disclosed via channel flow DNS that the footprint of outer layer VLSMs can be found even in the wall-shear stress fluctuations at high Reynolds numbers.

Lee and Sung (2013) compared the structure of TBL and turbulent pipe flow by means of DNSs and realised that VLSMs contribute more to the Reynolds shear stress in turbulent pipe flow than in TBL. In a subsequent numerical investigation, Lee et al. (2015) compared VLSMs in turbulent pipe and channel flows. They discovered that the population density of VLSMs and their lifetime in the core of the pipe is smaller than in the channel flow's corresponding location. Recently, Ahn and Sung (2017) have investigated the relation between the streamwise (ℓ_z) and the azimuthal length scale (ℓ_φ) for both LSMs and VLSMs in turbulent pipe flow at $\mathrm{Re}_\tau = 3008$. From two-point velocity correlations and one-dimensional pre-multiplied energy spectra, Ahn and Sung (2017) reached a ratio of $\ell_\varphi/\ell_z \approx 0.35$ for LSM and $\ell_\varphi/\ell_z \approx 0.07$ for VLSM.

Velocity Spikes

Even though there is a wide overall agreement of turbulence statistics obtained from DNSs and experimental investigations of wall-bounded turbulence, a discrepancy remains in the wall-normal velocity flatness near the wall. The high level of wall-normal flatness obtained from DNSs is usually missing in experimental data such as near-wall LDA measurements by Durst et al. (1995a). Xu et al. (1996) related this high flatness level to very large, local, wall-normal velocity fluctuations, which seldom appear in space and time in the vicinity of the wall, and termed them as velocity spikes. The authors of the latter study further compared DNS data with detailed near-wall LDA measurements and concluded that the above-mentioned extreme events are often not included in experimental data, which is due to limited time series, where extreme events are treated as outliers, a restricted access of optical measuring methods to the wall, or low measuring time resolutions. Nonetheless, den Toonder and Nieuwstadt (1997) measured higher flatness values of $F(u_r) \approx 20$ for $\mathrm{Re}_\tau = 520$ and $\mathrm{Re}_\tau = 690$ in turbulent pipe flow. However, their data exhibited scattering with respect to lower Reynolds number experiments of $\mathrm{Re}_\tau = 169$ and $\mathrm{Re}_\tau = 315$, where the wall-normal flatness did not exceed values of $F(u_r) = 8$. With regard to numerical investigations, Xu et al. (1996) reported $F(v) = 25$ for a turbulent plane channel flow at $\mathrm{Re}_\tau = 172$. Later, Lenaers et al. (2012) reported values between $F(v) = 27$ at $\mathrm{Re}_\tau = 180$ and $F(v) = 43$ at $\mathrm{Re}_\tau = 1000$ for the same type of flow. For numerical investigations of turbulent pipe flow, values of $F(u_r) = 19$ were reported by Eggels et al. (1994), $F(u_r) = 30$ by Wagner and Friedrich (1998), and $F(u_r) = 27$ or $F(u_r) = 31$ depending on the grid resolution by Shishkina and Wagner (2007), all of them at $\mathrm{Re}_\tau = 180$. Additionally, Boersma (2013) reported values of $F(u_r) = 31$ for $\mathrm{Re}_\tau = 688$ and $F(u_r) = 38$ for $\mathrm{Re}_\tau = 1842$. The scattering of the above-described wall-normal flatness values requires a revisit of high-order statistics in turbulent pipe flow, see chapter 4. The wall-normal flatness value in the vicinity of the wall is likely to converge very slowly with respect to the sampling time, since it reflects spatially- and temporarily-rare events. In particular, the dependence of the wall-normal flatness in the vicinity of the wall on the wall-normal grid spacing and on the averaging interval is studied in section 4.2.1 before its dependence on the Reynolds number is investigated in section 4.2.2. Subsequently, the dynamics of velocity spikes as well as their interplay with other coherent structures are discussed in section 4.3.

1.2.6 Inter-scale Energy Flux

Recalling the classical energy cascade of Richardson (1922) — introduced in section 1.2.4 —, turbulent kinetic energy is supposed to be transferred from large towards small scales, which is also referred to as *forward cascade*. However, it has been shown that energy can also be transferred in the opposite direction, i.e. from small towards large scales, which is denoted as *backscattering* or *inverse energy cascade*. As emphasised by Kraichnan (1967), von Kameke et al. (2011), and others, the inverse cascade is a key characteristic of two-dimensional turbulence[9]. Consequently, inverse cascades have been found in flow types, which resemble 2D flow dynamics such as rotating homogeneous turbulence (Pestana and Hickel, 2019). For wall-bounded flows, where the flow geometry is constraint in wall-normal direction, Härtel et al. (1994) tracked strong forward- and backscatter events related to strong shear layers near the wall. Correlating the inter-scale energy flux with the velocity field, Piomelli et al. (1996) noted that backscattering in the near-wall region is related to high-speed fluid sweeping toward the wall, whereas forward scattering is related to low-speed ejections. However, since the latter DNS studies of turbulent pipe and channel flow were restricted to Reynolds numbers as low as $\mathrm{Re}_\tau = 180$ and numerical domains not exceeding a length of $L_x \leq 4\pi h$, they were not able to observe inter-scale energy transfer related to LSMs and VLSMs. In this regard, recent studies by Cimarelli and fellow researchers (Cimarelli and De Angelis, 2011, 2012, 2014; Cimarelli et al., 2013, 2016, 2015) took the next steps by taking advantage of the space-scale duality of the generalised Kolmogorov equation of the second-order structure function introduced by Hill (2002). Cimarelli et al. (2016) discovered two different mechanisms of inter-scale energy transfer in wall-bounded turbulence. Taking the combined phase space of spatial and scale energy fluxes into account, they revealed an attached and a detached energy cascade, both of them involving backscattering of energy towards large scales away from the wall. Notwithstanding, since no direct structural information in terms of the scales responsible for the inverse energy cascade is obtained with their method, particular coherent structures cannot be linked with inter-scale energy flux events easily. Recently, Cho et al. (2018) and Lee and Moser (2019) have studied spectral energy budgets in turbulent plane channel flow, respectively at $\mathrm{Re}_\tau = 1700$ and $\mathrm{Re}_\tau = 5200$. Since the former study focused on spanwise scales only, they did not explicitly distinguish between LSMs and VLSMs, both of them corresponding to similar spanwise wavelengths. Cho et al. (2018) remarked that TKE production in the logarithmic layer occurs at spanwise scales proportional to the wall distance, whereas turbulent dissipation is related to spanwise wavelengths proportional to the Kolmogorov length scale. Besides the classical forward energy cascade, which they identified as the main energy transfer mechanism, Cho et al. (2018) spotted an inverse cascade in both the streamwise and spanwise velocity component. The study by Lee and Moser (2019), on the other hand, considered two-dimensional spectra in stream- and spanwise direction. Taking into account streamwise wave modes, they were able to observe TKE production at scales as large as VLSMs, while for spanwise scales they confirmed the findings of Cho et al. (2018). Despite of the studies mentioned

[9] For a comprehensive overview of two-dimensional turbulence, refer to Boffetta and Ecke (2012).

above, inter-scale energy transfer mechanisms related to VLSMs — particularly in turbulent pipe flow — are still widely unknown. Hence, the analysis in chapter 5 of this work elaborates on this topic, especially on the structural behaviour of inter-scale energy fluxes and on large-scale velocity fluctuations such as VLSMs.

1.3 Goals and Outline

Even though there is a good number of studies regarding coherent structures in turbulent flows in general and in turbulent pipe flow in specific, many of the dynamics and energy transfer mechanisms of the key coherent structures are still unexplored. While the near-wall cycle has been studied in great detail, open questions regarding the development and sustainment of LSMs and VLSMs in the logarithmic layer as well as their interaction with the near-wall cycle remain unanswered.

First of all, to settle, VLSMs require long computational domains. Hence, in a preliminary study, the question

1. Which computational domain length L is required to study the dynamics of VLSMs in turbulent pipe flow DNS?

will be addressed. The Reynolds number dependency of the streamwise Reynolds stress peak in the near-wall region of turbulent plane channel and of TBL flow has been described by several authors and linked to the interaction of large outer-flow motions with the wall (Hoyas and Jiménez, 2006; Hutchins and Marusic, 2007a). However, to the knowledge of the author, no study of the scaling behaviour of high-order statistics in turbulent pipe flow has been carried out so far. Therefore, one aim of this thesis is to determine the scaling behaviour of high-order statistics in turbulent pipe flow and link it to the interaction of turbulent coherent structures. However, since particularly high-order statistical moments require large averaging intervals to converge, a second preliminary study will be carried out to answer the following question:

2. Which averaging intervals are required for high-order moments of the velocity distribution in turbulent pipe flow to converge?

Subsequently, the scaling of turbulence statistics will be evaluated addressing the questions

3. How do high-order moments of the velocity distribution in turbulent pipe flow depend on the Reynolds number? Can this behaviour be approximated by a scaling law?

4. Is the scaling behaviour of high-order statistics related to the interaction of turbulent coherent structures, and if so, what is the underlying mechanism?

Regarding the near-wall cycle, it is widely accepted that SSMs in the fluctuating velocity field gain their energy from the mean field via the production mechanism, which is largest around $y^+ \approx 12$. In addition, there are forward and backward cascades of turbulent kinetic energy related to SSMs (see section 1.2.6). With respect to VLSMs, the mechanism, which provides them with energy, and the way they exchange energy with other coherent structures have not been unveiled, yet. Consequently, the questions

5. Which mechanism provides energy for VLSMs? Is there backscattering of energy towards VLSMs?

6. How large are the contributions of the transport terms and the dissipation term to the energy balance of VLSMs?

7. Does the direction of inter-scale energy transfer towards or away from VLSMs correlate with the moving direction of the latter (as it does for SSMs)?

8. How do the other terms of the VLSM budget equation correlate with the velocity field? What are the underlying coherent structure dynamics?

are addressed in this thesis. To answer these questions, DNSs of different Reynolds numbers Re_τ, different computational domain lengths L, and different near-wall wall-normal grid spacings Δr_{min} are to be carried out. While most one-point turbulence statistics, velocity spectra, and velocity correlations — quantities presented in chapters 3 and 4 — are computed on-the-fly during the runtime of the simulations, the terms of the scale-energy budget (chapter 5) are obtained from filtered instantaneous flow field realisations a posteriori.

The remaining part of this thesis is structured as follows. First, the finite-volume DNS method used to generate turbulent pipe flow data is described in chapter 2. The results section comprises chapters 3 to 5. Chapter 3 tackles the domain length dependency of turbulence statistics before the scaling and interaction of coherent structures in turbulent pipe flow are discussed by means of instantaneous observations as well as one-point and two-point turbulence statistics in chapter 4. As the last part of the results section, the energy transfer mechanisms of VLSM in turbulent pipe flow are analysed via the streamwise Reynolds stress budget equation of the low-pass filtered velocity field in chapter 5. Finally, an overall summary and conclusion is provided in chapter 6.

Chapter 2

Numerical Methodology

In order to solve the governing flow equations (1.8a) and (1.8b), a finite-volume method, based on the work of Schumann (1973, 1975) and Schumann et al. (1979), is applied. The method was implemented into the Fortran-based numerical solver FLOWSI (Schmitt and Friedrich, 1982, 1988; Schmitt et al., 1986), and further developed by Shishkina and Wagner (2004, 2005, 2007), and Feldmann and Wagner (2012) for turbulent pipe flow. The numerical solver FLOWSI has shown its validity to various different applications, ranging from DNS and LES of turbulent plane-channel (Grötzbach and Schumann, 1979; Schmitt et al., 1986) and pipe flow (Unger, 1994; Wagner, 1995; Wagner et al., 2001) to the simulation of thermal convection in different geometries (Bailon-Cuba et al., 2012; Horn et al., 2013; Kaczorowski and Wagner, 2009; Shishkina et al., 2009; Shishkina and Wagner, 2005; Wagner and Shishkina, 2013). The code has been parallelised by Feldmann (2015) for the case of turbulent flows through annular pipes with periodic boundaries in the axial and azimuthal direction. Within this work the code was further optimised and extended about several evaluations, i.e. one- and two-point turbulence statistics, spectral analysis and filtering methods. These techniques are described in chapters 3, 4, and 5 in more detail. In the following, the numerical method is described by means of the spatial and temporal discretisation (section 2.1), the implementation of the Poisson solver (section 2.2), and the computational parallelisation (section 2.3).

2.1 Discretisation

To solve the governing equations (1.8a) and (1.8b) numerically, they are discretised both in space and time.

2.1.1 Spatial Discretisation

The spatial discretisation involves the decomposition of the flow domain — depicted in figure 1.5 — into $N_z \times N_\varphi \times N_r$ finite-volume cells. In the two homogeneous directions (axial and azimuthal) an equidistant grid is applied in cylindrical coordinates. The axial and azimuthal coordinate of a finite-volume cell centred around (z_i, φ_j, r_k) are defined as

$$z_i = i\frac{L}{N_z} = i\Delta z, \quad \text{for } i = 1, \ldots, N_z, \tag{2.1}$$

$$\varphi_j = j\frac{2\pi}{N_\varphi} - \pi = j\Delta\varphi - \pi, \quad \text{for } j = 1, \ldots, N_\varphi, \tag{2.2}$$

where Δz and $\Delta\varphi$ are the sizes of the grid cell with respect to the axial and azimuthal direction, respectively. In wall-bounded turbulence, high velocity gradients occur in the wall-normal direction in the vicinity of the solid boundary, and thus, require near-wall refinement of the computational grid. A common approach for finite-difference and finite-volume schemes is a grid refinement based on the *tangens hyperbolicus* function as used by Moin and Kim (1982). FLOWSI features a *tangens hyperbolicus* based grid refinement, which has been introduced by Vinokur (1983). In this framework the radial grid spacing is given as an input parameter both at the wall (Δr_{N_r}) and the axis (Δr_1) of the cylindrical domain. The radial coordinate of a finite-volume cells centre and the radial size of the cell are then computed as follows:

$$r_k = \frac{1}{R}\left(\frac{\xi_k - \xi_{k-1}}{2} + \xi_{k-1}\right), \quad \text{for } k = 1, \ldots, N_r, \tag{2.3}$$

$$\Delta r_k = \frac{1}{R}\frac{\xi_k - \xi_{k-1}}{2}, \quad \text{for } k = 1, \ldots, N_r, \tag{2.4}$$

where

$$\xi_k = \frac{\eta_k}{\sqrt{\frac{\Delta r_{N_r}}{\Delta r_1}} + \left(1 - \sqrt{\frac{\Delta r_{N_r}}{\Delta r_1}}\right)\eta_k}, \quad \text{for } k = 0, \ldots, N_r, \tag{2.5}$$

and

$$\eta_k = \frac{1}{2}\left(1 + \frac{\tanh\left(\delta\left(\frac{k}{N_r} - \frac{1}{2}\right)\right)}{\tanh\left(\frac{\delta}{2}\right)}\right), \quad \text{for } k = 0, \ldots, N_r. \tag{2.6}$$

Here, δ is the solution of the iteratively solved non-linear equation

$$\frac{\sinh(\delta)}{\delta} = \frac{R}{N_r\sqrt{\Delta r_1 \Delta r_{N_r}}}. \tag{2.7}$$

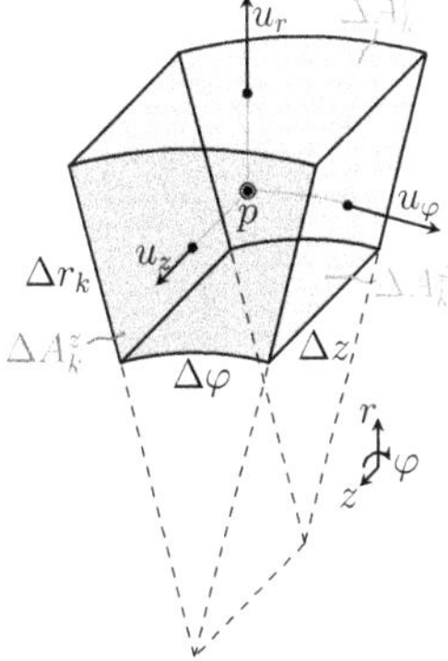

Figure 2.1: Finite-volume cell of the p-grid.

To avoid the decoupling of the primitive variables a staggered grid is introduced. Here, the pressure p and each velocity component (u_z,u_φ,u_r) are computed on a different grid. Figure 2.1 exemplarily depicts a finite volume V of the p-grid with a cell centre (z_i,φ_j,r_k). While the pressure is computed in the centre of the volume cell, the velocity components are evaluated on its boundaries. The surfaces bounding the finite-volume cell are computed as follows

$$\Delta A_k^z = \frac{\Delta\varphi}{2}\left(r_k+\frac{\Delta r_k}{2}\right)^2-\frac{\Delta\varphi}{2}\left(r_k-\frac{\Delta r_k}{2}\right)^2=\frac{1}{2}r_k\Delta\varphi\Delta r_k, \tag{2.8}$$

$$\Delta A_k^\varphi = \Delta z\Delta r_k, \tag{2.9}$$

$$\Delta A_k^r = \Delta z\left(r_k+\frac{\Delta r_k}{2}\right)\Delta\varphi, \tag{2.10}$$

$$\Delta A_{k-1}^r = \Delta z\left(r_k-\frac{\Delta r_k}{2}\right)\Delta\varphi, \tag{2.11}$$

for $k=1,\dots,N_r$, and the volume of the cell is given by

$$\Delta V_k = r_k\Delta\varphi\Delta z\Delta r_k, \text{ for } k=1,\dots,N_r. \tag{2.12}$$

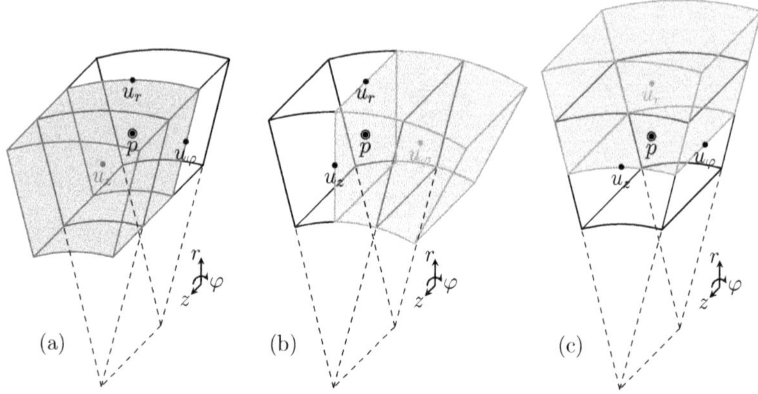

Figure 2.2: Finite-volume cells of the velocity grids V_α with respect to the p-grid. (a) V_z, (b) V_φ, (c) V_r.

Averages of a quantity $\phi \in \{p, u_z, u_\varphi, u_r\}$ over the bounding surfaces are defined as

$$ {}^{z\pm}\overline{\phi}(i,j,k) = \frac{1}{\Delta A_k^z} \int_{\varphi_j-\frac{\Delta\varphi}{2}}^{\varphi_j+\frac{\Delta\varphi}{2}} \int_{r_k-\frac{\Delta r_k}{2}}^{r_k+\frac{\Delta r_k}{2}} \phi(z_i \pm \Delta z/2, \varphi, r) r \,\mathrm{d}r\,\mathrm{d}\varphi, \tag{2.13}$$

$$ {}^{\varphi\pm}\overline{\phi}(i,j,k) = \frac{1}{\Delta A_k^\varphi} \int_{z_i-\frac{\Delta z}{2}}^{z_i+\frac{\Delta z}{2}} \int_{r_k-\frac{\Delta r_k}{2}}^{r_k+\frac{\Delta r_k}{2}} \phi(z, \varphi_j \pm \Delta\varphi/2, r) \,\mathrm{d}r\,\mathrm{d}z, \tag{2.14}$$

$$ {}^{r+}\overline{\phi}(i,j,k) = \frac{1}{\Delta A_k^r} \int_{z_i-\frac{\Delta z}{2}}^{z_i+\frac{\Delta z}{2}} \int_{\varphi_j-\frac{\Delta\varphi}{2}}^{\varphi_j+\frac{\Delta\varphi}{2}} \phi(z, \varphi, r_k + \Delta r_k/2)(r_k + \Delta r_k/2) \,\mathrm{d}\varphi\,\mathrm{d}z, \tag{2.15}$$

$$ {}^{r-}\overline{\phi}(i,j,k) = \frac{1}{\Delta A_{k-1}^r} \int_{z_i-\frac{\Delta z}{2}}^{z_i+\frac{\Delta z}{2}} \int_{\varphi_j-\frac{\Delta\varphi}{2}}^{\varphi_j+\frac{\Delta\varphi}{2}} \phi(z, \varphi, r_k - \Delta r_k/2)(r_k - \Delta r_k/2) \,\mathrm{d}\varphi\,\mathrm{d}z, \tag{2.16}$$

and the volume average over the cell volume ΔV_k reads

$$ \overline{\phi}(i,j,k) = \frac{1}{\Delta V_k} \int_{z_i-\frac{\Delta z}{2}}^{z_i+\frac{\Delta z}{2}} \int_{\varphi_j-\frac{\Delta\varphi}{2}}^{\varphi_j+\frac{\Delta\varphi}{2}} \int_{r_k-\frac{\Delta r_k}{2}}^{r_k+\frac{\Delta r_k}{2}} \phi(z, \varphi, r) r \,\mathrm{d}r\,\mathrm{d}\varphi\,\mathrm{d}z. \tag{2.17}$$

With respect to the pressure grid, the u_z-, u_φ- and u_r-grids are shifted by $\Delta z/2$, $\Delta\varphi/2$, and $\Delta r_k/2$, respectively (figure 2.2). Consequently, the centre of the shifted volume V_α with $\alpha \in \{z, \varphi, r\}$ coincides with the corresponding bounding surface of the pressure volume V. While the volume of the cells shifted in the homogeneous directions is congruent to the volume of the corresponding pressure cell — $\Delta V_{z,k} = \Delta V_{\varphi,k} = \Delta V_k$, and $\Delta A_{\alpha,k}^\beta = \Delta A_k^\beta$ for $\alpha \in \{z, \varphi\}$ the grid, and $\beta \in \{z, \varphi, r\}$ the direction of the surfaces normal vector —, the volume of the cell shifted in the radial direction is not congruent to the volume of the pressure cell — $\Delta V_{r,k} \neq \Delta V_k$, $\Delta A^{\beta}_{r,k} \neq \Delta A^{\beta}_{k}$ for $\alpha \in \{z,\varphi,r\}$. In the following, $\overline{p}$ denotes the pressure averaged over ΔV_k, while $\overline{u_\alpha}$ denotes the velocity component averaged over the corresponding finite volume $\Delta V_{\alpha,k}$ with $\alpha \in \{z,\varphi,r\}$. Additionally, ${}^{\beta\pm}\overline{\phi}$ indicates surface averaging of a quantity ϕ over the surfaces bounding ΔV_k normal to β, and ${}^{\alpha\beta\pm}\overline{\phi}$ indicates surface averaging of a quantity ϕ over the surfaces bounding $\Delta V_{\alpha,k}$ normal to β, where $\{\alpha,\beta\} \in \{z,\varphi,r\}$. Integrating equation (1.8a) over ΔV_k results in the finite-volume representation of the continuity equation

$$\left({}^{z+}\overline{u_z} - {}^{z-}\overline{u_z}\right)\Delta A^z_k + \left({}^{\varphi+}\overline{u_\varphi} - {}^{\varphi-}\overline{u_\varphi}\right)\Delta A^\varphi_k + \left({}^{r+}\overline{u_r}\Delta A^r_k - {}^{r-}\overline{u_r}\Delta A^r_{k-1}\right) = 0, \quad (2.18)$$

where advantage is taken of the fact that the surface averaged velocities on the boundaries of the pressure volume cell V can be approximated with the volume averaged velocities of the velocity volume cells V_α, i.e. ${}^{\beta+}\overline{u_\alpha} \approx \overline{u_\alpha}$. Integrating the momentum equations (1.8b) over the corresponding velocity volume cell V_α gives the finite-volume representation of the momentum equations as follows

$$\frac{\partial \overline{u_\alpha}}{\partial t} + K_\alpha - D_\alpha + P_\alpha - C_\alpha = 0, \text{ for } \alpha = z,\varphi,r, \quad (2.19)$$

where

$$\begin{aligned} K_\alpha &= \frac{\Delta A^z_{\alpha,k}}{\Delta V_{\alpha,k}}\left({}^{\alpha z+}\overline{u_\alpha}\cdot{}^{\alpha z+}\overline{u_z} - {}^{\alpha z-}\overline{u_\alpha}\cdot{}^{\alpha z-}\overline{u_z}\right) \\ &+ \frac{\Delta A^\varphi_{\alpha,k}}{\Delta V_{\alpha,k}}\left({}^{\alpha\varphi+}\overline{u_\alpha}\cdot{}^{\alpha\varphi+}\overline{u_\varphi} - {}^{\alpha\varphi-}\overline{u_\alpha}\cdot{}^{\alpha\varphi-}\overline{u_\varphi}\right) \\ &+ \frac{1}{\Delta V_{\alpha,k}}\left(\Delta A^r_{\alpha,k}\cdot{}^{\alpha r+}\overline{u_\alpha}\cdot{}^{\alpha r+}\overline{u_r} - \Delta A^r_{\alpha,k-1}\cdot{}^{\alpha r-}\overline{u_\alpha}\cdot{}^{\alpha r-}\overline{u_r}\right) \end{aligned} \quad (2.20)$$

is the convective transport term and

$$\begin{aligned} D_\alpha &= \frac{2\Delta A^z_{\alpha,k}}{\mathrm{Re}_\tau \Delta V_{\alpha,k}}\left({}^{\alpha z+}\overline{S_{\alpha z}} - {}^{\alpha z-}\overline{S_{\alpha z}}\right) + \frac{2\Delta A^\varphi_{\alpha,k}}{\mathrm{Re}_\tau \Delta V_{\alpha,k}}\left({}^{\alpha\varphi+}\overline{S_{\alpha\varphi}} - {}^{\alpha\varphi-}\overline{S_{\alpha\varphi}}\right) \\ &+ \frac{2}{\mathrm{Re}_\tau \Delta V_{\alpha,k}}\left(\Delta A^r_{\alpha,k}\cdot{}^{\alpha r+}\overline{S_{\alpha r}} - \Delta A^r_{\alpha,k-1}\cdot{}^{\alpha r-}\overline{S_{\alpha r}}\right) \end{aligned} \quad (2.21)$$

is the diffusive transport term with the surface averaged components of the rate-of-strain tensor ${}^{\alpha\beta\pm}\overline{S_{\alpha\beta}}$

$$
\begin{aligned}
{}^{zz\pm}\overline{S_{zz}} &= {}^{zz\pm}\overline{\frac{\partial u_z}{\partial z}}, && \text{(2.22a)}\\
{}^{z\varphi\pm}\overline{S_{z\varphi}} &= \frac{1}{2}\left(\frac{1}{r}\,{}^{z\varphi\pm}\overline{\frac{\partial u_z}{\partial \varphi}} + {}^{z\varphi\pm}\overline{\frac{\partial u_\varphi}{\partial z}}\right), && \text{(2.22b)}\\
{}^{zr\pm}\overline{S_{zr}} &= \frac{1}{2}\left({}^{zr\pm}\overline{\frac{\partial u_z}{\partial r}} + {}^{zr\pm}\overline{\frac{\partial u_r}{\partial z}}\right), && \text{(2.22c)}\\
{}^{\varphi z\pm}\overline{S_{\varphi z}} &= \frac{1}{2}\left({}^{\varphi z\pm}\overline{\frac{\partial u_\varphi}{\partial z}} + \frac{1}{r}\,{}^{\varphi z\pm}\overline{\frac{\partial u_z}{\partial \varphi}}\right), && \text{(2.22d)}\\
{}^{\varphi\varphi\pm}\overline{S_{\varphi\varphi}} &= \frac{1}{r}\left({}^{\varphi\varphi\pm}\overline{\frac{\partial u_\varphi}{\partial \varphi}} + {}^{\varphi\varphi\pm}\overline{u_r}\right), && \text{(2.22e)}\\
{}^{\varphi r\pm}\overline{S_{\varphi r}} &= \frac{1}{2}\left({}^{\varphi r\pm}\overline{\frac{\partial u_\varphi}{\partial r}} + \frac{1}{r}\,{}^{\varphi r\pm}\overline{\frac{\partial u_r}{\partial \varphi}} - \frac{1}{r}\,{}^{\varphi r\pm}\overline{u_\varphi}\right), && \text{(2.22f)}\\
{}^{rz\pm}\overline{S_{rz}} &= \frac{1}{2}\left({}^{rz\pm}\overline{\frac{\partial u_r}{\partial z}} + {}^{rz\pm}\overline{\frac{\partial u_z}{\partial r}}\right), && \text{(2.22g)}\\
{}^{r\varphi\pm}\overline{S_{r\varphi}} &= \frac{1}{2}\left(\frac{1}{r}\,{}^{r\varphi\pm}\overline{\frac{\partial u_r}{\partial \varphi}} - \frac{1}{r}\,{}^{r\varphi\pm}\overline{u_\varphi} + {}^{r\varphi\pm}\overline{\frac{\partial u_\varphi}{\partial r}}\right), && \text{(2.22h)}\\
{}^{rr\pm}\overline{S_{rr}} &= {}^{rr\pm}\overline{\frac{\partial u_r}{\partial r}}. && \text{(2.22i)}
\end{aligned}
$$

The pressure term reads

$$
\begin{aligned}
P_z &= \frac{\Delta A^z_{z,k}}{\Delta V_{z,k}}\left({}^{zz+}\overline{p} - {}^{zz-}\overline{p}\right), && \text{(2.23)}\\
P_\varphi &= \frac{\Delta A^\varphi_{\varphi,k}}{\Delta V_{\varphi,k}}\left({}^{\varphi\varphi+}\overline{p} - {}^{\varphi\varphi-}\overline{p}\right), && \text{(2.24)}\\
P_r &= \frac{1}{\Delta V_{r,k}}\left(\Delta A^r_{r,k}\cdot{}^{rr+}\overline{p} - \Delta A^r_{r,k-1}\cdot{}^{rr-}\overline{p}\right), && \text{(2.25)}
\end{aligned}
$$

and the curvature term is denoted as

$$C_z = 0, \tag{2.26}$$

$$C_\varphi = \frac{\Delta\varphi \cdot \Delta A_{\varphi,k}}{\Delta V_{\varphi,k}} \left(\frac{2}{\mathrm{Re}_\tau} \cdot {}^{\varphi}\overline{S_{\varphi r}} - {}^{\varphi}\overline{u_\varphi} \cdot {}^{\varphi}\overline{u_r} \right), \tag{2.27}$$

$$C_r = \frac{\Delta\varphi \cdot \Delta A_{\varphi,k}}{\Delta V_{r,k}} \left({}^{\varphi}\overline{p} + {}^{\varphi}\overline{u_\varphi}^2 - \frac{2}{\mathrm{Re}_\tau} \cdot {}^{\varphi}\overline{S_{\varphi\varphi}} \right). \tag{2.28}$$

All surface averaged quantities, which are not already defined by volume averages of cell centres are approximated by means of a fourth-order accurate polynomial approximation scheme — further described in Shishkina and Wagner (2007).

2.1.2 Temporal Discretisation

The **explicit Leapfrog-Euler time integration scheme** — introduced by Schumann (1973) — for equation (2.19) yields

$$\frac{\overline{u_\alpha^{n+1}} - f_1 \overline{u^n} - f_2 \overline{u_\alpha^{n-1}}}{f_3 \Delta t} + K_\alpha^n - D_\alpha^{n-1+f_4} + P_\alpha^n - C_\alpha^n = 0, \text{ for } \alpha = z, \varphi, r, \tag{2.29}$$

where n is the number of the time step of the evaluated flow quantity, Δt is the temporal increment between two time steps and f_i with $i \in 1,..,4$ are prefactors that depend on the stage of the integration cycle. Each cycle — consisting of $N = 50$ time steps typically — begins with an Euler step ($n = 1$), is then followed by $N - 2$ Leapfrog steps and is closed by a combined Leapfrog-Euler step ($n = N$). The corresponding prefactors for each stage of the integration scheme are given in table 2.1. The Leapfrog-Euler

Table 2.1: Prefactors for the explicit Leapfrog-Euler time integration scheme as introduced by Schumann (1973).

scheme	time step	f_1	f_2	f_3	f_4
Euler	$n = 1$	1	0	1	1
Leapfrog	$1 < n < N$	0	1	2	0
Euler-Leapfrog	$n = N$	$\frac{1}{2}$	$\frac{1}{2}$	$\frac{3}{2}$	0

scheme is — unlike the classic Leapfrog scheme — stable and of second-order accuracy. Shishkina and Wagner (2007) derive the von Neumann stability criterion of the explicit

Leapfrog-Euler scheme as

$$\Delta t_{exp}^{crit} < \left(\frac{3}{2} \sum_{\alpha=z,\varphi,r} \frac{U_\alpha}{\Delta x_\alpha} + \frac{16}{3} \frac{1}{\mathrm{Re}_\tau} \sum_{\alpha=z,\varphi,r} \frac{1}{\Delta x_\alpha^2} \right)^{-1}, \tag{2.30}$$

where Δt_{exp}^{crit} is an estimate for the critical time step, $\Delta x_\alpha = \Delta\alpha$ for $\alpha = z, r$ and $\Delta x_r = r\Delta\varphi$ are the mesh sizes in each spatial direction, and U_α is the velocity component, which is considered to be constant in the von Neumann stability analysis. Since the von Neumann stability analysis is only valid for linear problems the critical time step is multiplied by safety factor of 0.5 to cover non-linear instabilities. Finally, the critical time step is computed as follows,

$$\begin{aligned} \Delta t_{exp}^{crit} = 0.5 \Bigg(\max_{z_i,\varphi_j,r_k} \Bigg\{ \frac{3}{2} \Bigg(& \frac{|\overline{u_z}(z_i+0.5\Delta z,\varphi_j,r_k)|}{\Delta z} + \frac{|\overline{u_\varphi}(z_i,\varphi_j+0.5\Delta\varphi,r_k)|}{r_k\Delta\varphi} \\ + & \frac{|\overline{u_r}(z_i,\varphi_j,r_k+0.5\Delta r_k)|}{\Delta r_k} \Bigg) + \frac{16}{3}\frac{1}{\mathrm{Re}_\tau} \left(\frac{1}{\Delta z^2} + \frac{1}{(r_k\Delta\varphi)^2} + \frac{1}{\Delta r_k^2} \right) \Bigg\} \Bigg)^{-1}. \end{aligned} \tag{2.31}$$

Due to the small size of the finite volumes in the azimuthal direction in the vicinity of the cylindrical axis, the stability criterion requires very small time steps for the explicit scheme. To overcome this problem an **implicit scheme** — as introduced by Kim and Moin (1985) — is applied to the azimuthal component of the convective term K_α and the diffusive term D_α in a subdomain around the axis. Consequently, the momentum transport equation results in

$$\begin{aligned} \frac{\overline{u_\alpha^{n+1}} - f_1\overline{u_\alpha^n} - f_2\overline{u_\alpha^{n-1}}}{f_3\Delta t} + K_{1,\alpha}^n + K_{2,\alpha}^{n+1,n} - D_{1,\alpha}^{n-1+f_4} - D_{2,\alpha}^{n-1+f_4,n+1} + P_\alpha^n - C_\alpha^n = 0, \\ \text{for } \alpha = z,\varphi,r, \end{aligned} \tag{2.32}$$

where

$$\begin{aligned} K_{1,\alpha}^n &= \frac{\Delta A_{\alpha,k}^z}{\Delta V_{\alpha,k}} \left({}^{\alpha z+}\overline{u_\alpha^n} \cdot {}^{\alpha z+}\overline{u_z^n} - {}^{\alpha z-}\overline{u_\alpha^n} \cdot {}^{\alpha z-}\overline{u_z^n} \right) \\ &+ \frac{1}{\Delta V_{\alpha,k}} \left(\Delta A_{\alpha,k}^r \cdot {}^{\alpha r+}\overline{u_\alpha^n} \cdot {}^{\alpha r+}\overline{u_r^n} - \Delta A_{\alpha,k-1}^r \cdot {}^{\alpha r-}\overline{u_\alpha^n} \cdot {}^{\alpha r-}\overline{u_r^n} \right), \end{aligned} \tag{2.33}$$

$$K_{2,\alpha}^{n,m} = \frac{\Delta A_{\alpha,k}^\varphi}{\Delta V_{\alpha,k}} \left({}^{\alpha\varphi+}\overline{u_\alpha^m} \cdot {}^{\alpha\varphi+}\overline{u_\varphi^n} - {}^{\alpha\varphi-}\overline{u_\alpha^m} \cdot {}^{\alpha\varphi-}\overline{u_\varphi^n} \right), \tag{2.34}$$

and

$$\begin{aligned} D_{1,\alpha}^{n} &= \frac{2\Delta A_{\alpha,k}^{z}}{\mathrm{Re}_\tau \Delta V_{\alpha,k}} \left({}^{\alpha z+}\overline{S_{\alpha z}^{n}} - {}^{\alpha z-}\overline{S_{\alpha z}^{n}} \right) \\ &+ \frac{2}{\mathrm{Re}_\tau \Delta V_{\alpha,k}} \left(\Delta A_{\alpha,k}^{r} \cdot {}^{\alpha r+}\overline{S_{\alpha r}^{n}} - \Delta A_{\alpha,k-1}^{r} \cdot {}^{\alpha r-}\overline{S_{\alpha r}^{n}} \right), \qquad (2.35) \\ D_{2,\alpha}^{n,m} &= +\frac{2\Delta A_{\alpha,k}^{\varphi}}{\mathrm{Re}_\tau \Delta V_{\alpha,k}} \left({}^{\alpha\varphi+}\overline{S_{\alpha\varphi}^{n,m}} - {}^{\alpha\varphi-}\overline{S_{\alpha\varphi}^{n,m}} \right). \qquad (2.36) \end{aligned}$$

The surface averaged rate-of-strain tensor of mixed time steps ${}^{\alpha\beta\pm}\overline{S_{\alpha\beta}^{n,m}}$ can be found in Shishkina and Wagner (2007). Analogously to the explicit time integration scheme, a critical time step for the semi-implicit scheme yields

$$\begin{aligned} \Delta t_{imp}^{crit} = 0.5 \Bigg(\max_{z_i,\varphi_j,r_k} \Bigg\{ \frac{3}{2} \left(\frac{|\overline{u_z}(z_i+0.5\Delta z,\varphi_j,r_k)|}{\Delta z} + \frac{|\overline{u_r}(z_i,\varphi_j,r_k+0.5\Delta r_k)|}{\Delta r_k} \right) \\ + \frac{16}{3}\frac{1}{\mathrm{Re}_\tau} \left(\frac{1}{\Delta z^2} + \frac{1}{\Delta r_k^2} \right) \Bigg\} \Bigg)^{-1}. \qquad (2.37) \end{aligned}$$

The radial border between the regions of explicit and semi-implicit time integration is computed in a way that the critical time step is as large as possible with the semi-implicit region being as small as possible. Further details are depicted in Shishkina and Wagner (2007).

2.2 Time Integration and Poisson Solver

The solution of equation (2.29) — or (2.32) respectively — involves a projection method first introduced by Chorin (1967, 1968). The algorithm consists of three basic steps and is in the following discussed by means of the explicit Leapfrog formulation of equation (2.29), i.e.

$$\frac{\overline{u_\alpha^{n+1}} - \overline{u_\alpha^{n-1}}}{2\Delta t} + K_\alpha^n - D_\alpha^{n-1} + P_\alpha^n - C_\alpha^n = 0, \text{ for } \alpha = z,\varphi,r. \qquad (2.38)$$

First, the pressure term is neglected from equation (2.38) and an auxiliary velocity field $\mathbf{u}^* = (u_z^*, u_\varphi^*, u_r^*)$ is computed from

$$\frac{\overline{u_\alpha^{*}} - \overline{u_\alpha^{n-1}}}{2\Delta t} + K_\alpha^n - D_\alpha^{n-1} - C_\alpha^n = 0, \text{ for } \alpha = z,\varphi,r. \qquad (2.39)$$

Then, the Poisson equation

$$\begin{aligned}
\nabla^2\overline{p^n} &= \frac{\nabla\cdot\overline{\mathbf{u}^*}}{2\Delta t} \\
&= \frac{1}{2\Delta t\cdot\Delta V_k}\Big[\left({}^{z+}\overline{u_z^*}-{}^{z-}\overline{u_z^*}\right)\Delta A_k^z+\left({}^{\varphi+}\overline{u_\varphi^*}-{}^{\varphi-}\overline{u_\varphi^*}\right)\Delta A_k^\varphi \\
&+ \left({}^{r+}\overline{u_r^*}\Delta A_k^r-{}^{r-}\overline{u_r^*}\Delta A_{k-1}^r\right)\Big]
\end{aligned} \tag{2.40}$$

is solved for the auxiliary velocity $\mathbf{u}^*$ with the boundary condition $(\mathbf{n}\cdot\nabla\overline{p^n})|_{r=R}=0$ at the solid wall ($\mathbf{n}$ is the normal vector). Finally, the velocity field at time step $n+1$ is updated from the auxiliary field and the solution of the Poisson equation

$$\overline{\mathbf{u}^{n+1}}=\overline{\mathbf{u}^*}-2\Delta t\cdot\nabla\overline{p^n}. \tag{2.41}$$

For the Leapfrog formulation of the semi-implicit equation (2.32),

$$\begin{aligned}
&\frac{\overline{u_\alpha^{n+1}}-\overline{u_\alpha^{n-1}}}{2\Delta t}+K_{1,\alpha}^n+K_{2,\alpha}^{n+1,n}-D_{1,\alpha}^{n-1}-D_{2,\alpha}^{n-1,n+1}+P_\alpha^n-C_\alpha^n=0, \\
&\qquad\text{for } \alpha=z,\varphi,r,
\end{aligned} \tag{2.42}$$

the auxiliary velocity field $\mathbf{u}^*$ is computed from

$$\begin{aligned}
&\frac{\overline{u_\alpha^*}-\overline{u_\alpha^{n-1}}}{2\Delta t}+K_{1,\alpha}^n+K_{2,\alpha}^{*,n}-D_{1,\alpha}^{n-1}-D_{2,\alpha}^{n-1,*}-C_\alpha^n=0, \\
&\qquad\text{for } \alpha=z,\varphi,r.
\end{aligned} \tag{2.43}$$

Now the Poisson equation is solved for an auxiliary function $\overline{\psi^n}$

$$\nabla^2\overline{\psi^n}=\nabla\cdot\overline{\mathbf{u}^*}, \tag{2.44}$$

with the boundary condition $(\mathbf{n}\cdot\nabla\overline{\psi^n})|_{r=R}=0$. Finally, velocity and pressure are updated as follows

$$\overline{\mathbf{u}^{n+1}} = \overline{\mathbf{u}^*}-\nabla\overline{\psi^n}, \tag{2.45}$$

$$\overline{p^n} = \left(\frac{1}{2\Delta t}+\overline{u_\varphi^n}\frac{\partial}{\partial\varphi}+\frac{\partial\overline{u_\varphi^n}}{\partial\varphi}-\frac{1}{\mathrm{Re}_\tau}\frac{\partial^2}{\partial\varphi^2}\right)\overline{\psi^n}. \tag{2.46}$$

2.3 Parallelisation

The DNS of high-Reynolds-number flow requires computational resources in terms of both computing power and memory. To distribute the computational resources to

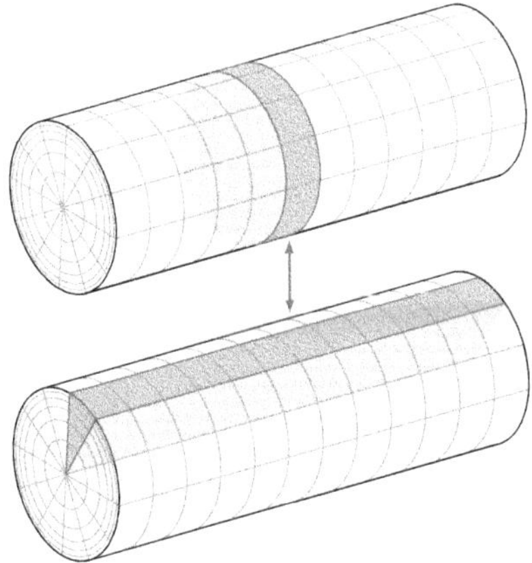

Figure 2.3: Domain decomposition strategy.

several processors and gain advantage of high performance computing (HPC) clusters, the current version of FLOWSI has been parallelised via the message passing interface (MPI). The parallelisation strategy involves the axial decomposition of the velocity field $\overline{\mathbf{u}}(z_i, \varphi_j, r_k)$ and the pressure field $\overline{p}(z_i, \varphi_j, r_k)$ into n_{proc} slices, where n_{proc} is the number of computational cores. Each slice contains then $N_z/n_{proc} \times N_\varphi \times N_r$ finite Volume cells. The solution of the Poisson equation (2.40) or (2.44) requires fast Fourier transforms (FFTs) of the velocity field with respect to the axial and the azimuthal direction in order to decouple the equation from these directions. Since for the FFT algorithm in axial direction the full axial signal is required on each core, the data — after being Fourier transformed in the azimuthal direction — is transposed from an axial to an azimuthal decomposition by means of a "shuffling" algorithm, as depicted in figure 2.3. After being Fourier transformed in axial and azimuthal direction the Poisson equation simplifies to a one-dimensional Helmholtz problem for each independent wave number pair and can, thus, be solved by means of a tridiagonal matrix algorithm. Thereafter, the solution of the equation is transformed back to physical space in the axial direction. Then, the solution is redistributed back from azimuthal decomposition to axial decomposition via the shuffling algorithm. Finally, the solution is Fourier transformed back to physical space in the azimuthal direction.

Chapter 3

Domain-length Dependency of Turbulent Pipe Flow

In a preliminary study, the influence of the computational domain length L on the flow structure and turbulence statistics is investigated. Since this work focuses on the largest turbulent scales of motion, it is of utmost importance to fully capture these scales in the computational domain. Hence, instantaneous flow field realisations, as well as turbulence one- and two-point statistics, are presented for $\mathrm{Re}_\tau = 180$ in computational domains of lengths $3.5R \leq L \leq 84R$ and for $\mathrm{Re}_\tau = 720$ in domains of length $3.5R \leq L \leq 42R$ in the following. An overview of the different setups used in this chapter is presented in table 3.1. Parts of the analysis have been published in Feldmann et al. (2018).

Table 3.1: Turbulent pipe-flow simulation cases. N_z, N_φ and N_r are the number of grid points with respect to the axial, azimuthal and radial direction, respectively. Δz^+, streamwise grid spacing; $R^+\Delta\varphi$, azimuthal grid spacing at the wall; Δr^+_{min} and Δr^+_{max}, minimal and maximal radial grid spacing; all grid are spacings normalised in wall units.

	P180L3	P180L7	P180L14	P180	P180L84	P720L3	P720L7	P720L14	P720
style	-○-	-○-	-◇-	-□-	-+-	-●-	-●-	-◆-	-△-
Re_τ	180	180	180	180	180	720	720	720	720
Re_b	5258	5258	5258	5258	5258	25 992	25 992	25 992	25 992
L/R	3.5	7	14	42	84	3.5	7	14	42
N_z	128	256	512	1536	3072	384	768	1536	4608
N_φ	256	256	256	256	256	1024	1024	1024	1024
N_r	84	84	84	84	84	222	222	222	222
Δz^+	4.9	4.9	4.9	4.9	4.9	6.6	6.6	6.6	6.6
$R^+\Delta\varphi$	4.4	4.4	4.4	4.4	4.4	4.4	4.4	4.4	4.4
Δr^+_{min}	0.31	0.31	0.31	0.31	0.31	0.49	0.49	0.49	0.49
Δr^+_{max}	4.4	4.4	4.4	4.4	4.4	6.6	6.6	6.6	6.6

Instantaneous representations of the fluctuating velocity components are presented in figures 3.1–3.3 for $\mathrm{Re}_\tau = 180$ and figures 3.4–3.6 for $\mathrm{Re}_\tau = 720$. For both Reynolds numbers, the streaky structures visible in the rz-plane in the streamwise fluctuating velocity (figure 3.1 and 3.4, respectively) are much longer in z-direction, and thus, more critical for small computational domain lengths than the other velocity components (figures 3.2, 3.3, 3.5, and 3.5). For the largest domain length considered ($L = 42R$) at $\mathrm{Re}_\tau = 180$, low- and high-speed streamwise velocity fluctuations of length $l_z > 7R$ manifest themselves in figure 3.1 at ($r/R \approx 0.8$, $-1 \gtrsim z/R \gtrsim 7$) and ($r/R \approx -0.9$, $36 \gtrsim z/R \gtrsim 44$. Note that both structures cross the boundaries of the periodic domain at $z/R = 0, 42$. Such long structures cannot be represented in the smaller domains of $L = 3.5R$ (figure 3.1a) and $L = 7R$ (figure 3.1b). The azimuthal and radial components of the fluctuating velocity at $\mathrm{Re}_\tau = 180$ shown in figure 3.2 and 3.2, on the contrary, are not longer than $3R$, and thus, fit in the smallest computational box of length $L = 3.5R$ (figures 3.2a and 3.3a).

For $\mathrm{Re}_\tau = 720$ the range of scales increases with respect to $\mathrm{Re}_\tau = 180$ and, as the iso-contours in the $r\varphi$-planes in figures 3.4–3.6 reflect, small viscous scales are located closer to the wall than for $\mathrm{Re}_\tau = 180$. In the bulk flow region, larger scales appear in all velocity components (figures 3.4–3.6), the largest of them in the streamwise component (figure 3.4). These streamwise elongated fluctuations of the streamwise velocity component u_z' are representatives of VLSMs. In figure 3.4(d) a high-speed VLSM of length $l_z \approx 10R$ is settled at ($r/R \approx -0.5$, $3 \gtrsim z/R \gtrsim 13$). As for $\mathrm{Re}_\tau = 180$, such long coherent structures cannot be obtained from computational domain lengths of $L = 3.5$ (figure 3.4a) and $L = 7$ (figure 3.4b). Coherent structures in the iso-contours of the azimuthal and radial velocity fluctuation, shown in rz-planes in figures 3.5 and 3.6, reach their maximum streamwise extent in the bulk flow region. Being shorter than $3R$, these structures, associated with LSMs, can be represented in computational boxes as short as $L/R = 3.5$ (figure 3.5a and 3.6a).

The inspection of instantaneous flow field realisations suggests computational domain lengths of $L = 14R$ to be sufficiently large to capture all relevant scales in the turbulent velocity field at the two friction Reynolds numbers of $\mathrm{Re}_\tau = 180$ and $\mathrm{Re}_\tau = 720$. However, due to the nature of turbulent flows, no general conclusions can be drawn from instantaneous snapshots and statistical analysis is required. Therefore, the domain-length dependency on both one- and two-point statistical quantities is determined in the following.

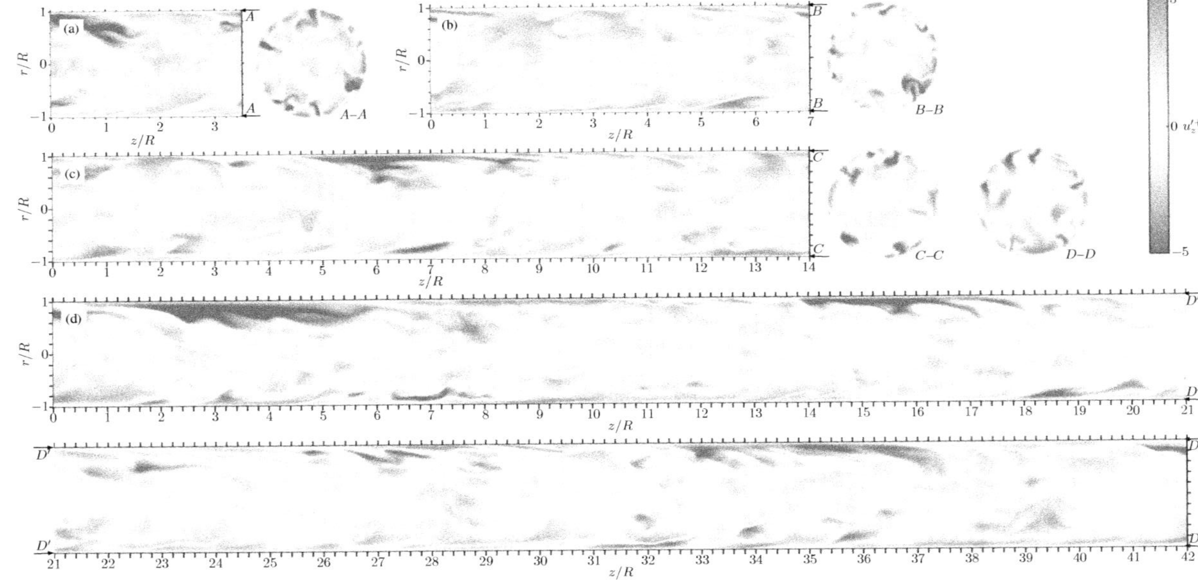

Figure 3.1: Instantaneous streamwise fluctuating velocity $u_z'^+$ for different L in rz- and $r\varphi$-planes at $\mathrm{Re}_\tau = 180$. (a) P180L3; (b) P180L7 (C) P180L14; (d) P180.

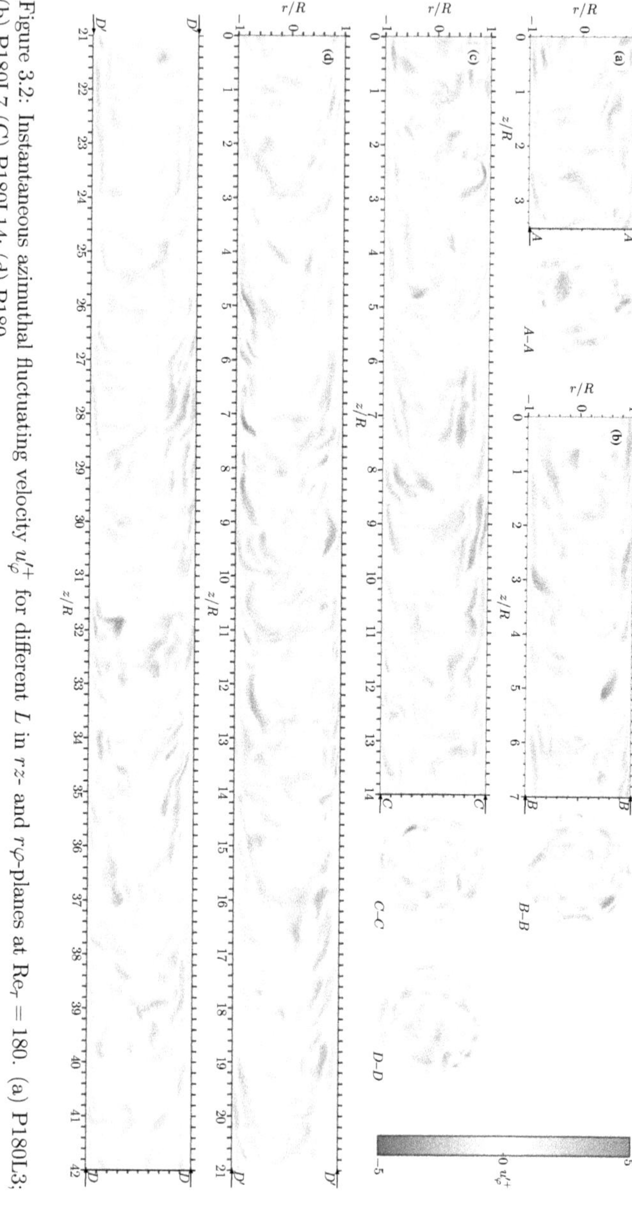

Figure 3.2: Instantaneous azimuthal fluctuating velocity u'^{+}_{φ} for different L in rz- and $r\varphi$-planes at $\mathrm{Re}_{\tau} = 180$. (a) P180L3; (b) P180L7 (C) P180L14; (d) P180.

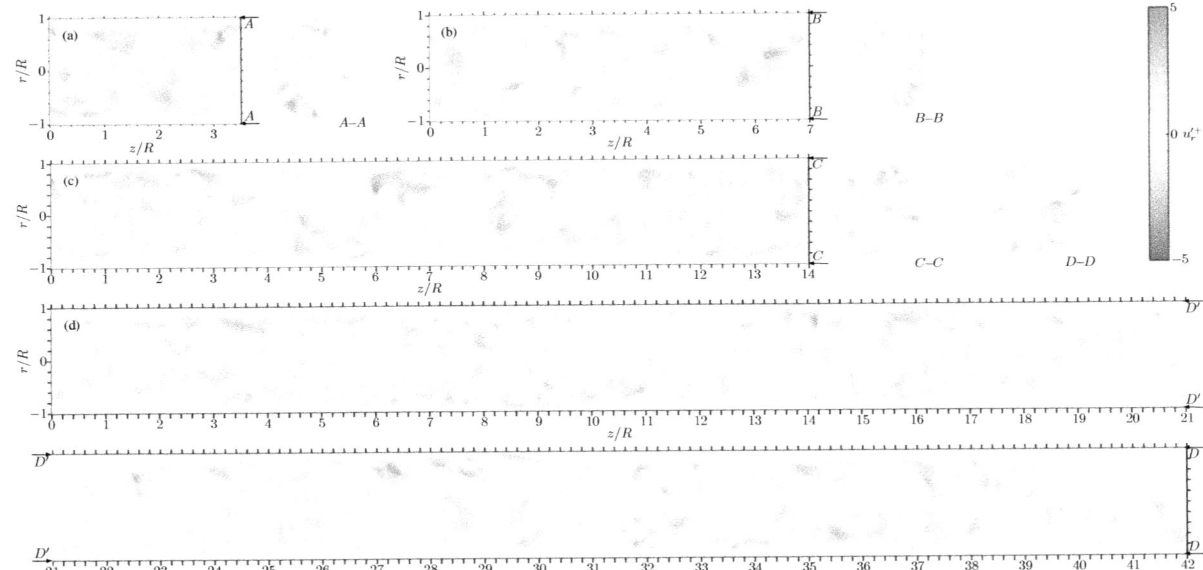

Figure 3.3: Instantaneous radial fluctuating velocity $u_r'^+$ for different L in rz- and $r\varphi$-planes at $\mathrm{Re}_\tau = 180$. (a) P180L3; (b) P180L7 (C) P180L14; (d) P180.

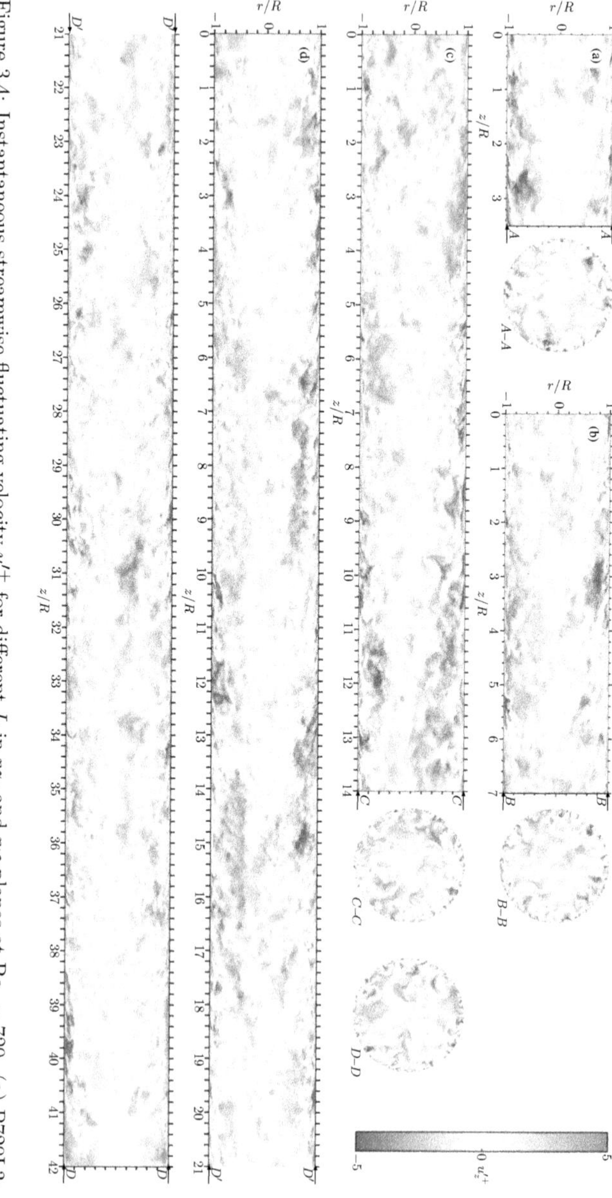

Figure 3.4: Instantaneous streamwise fluctuating velocity $u_z'^+$ for different L in rz- and $r\varphi$-planes at $\mathrm{Re}_\tau = 720$. (a) P720L3; (b) P720L7 (C) P720L14; (d) P720.

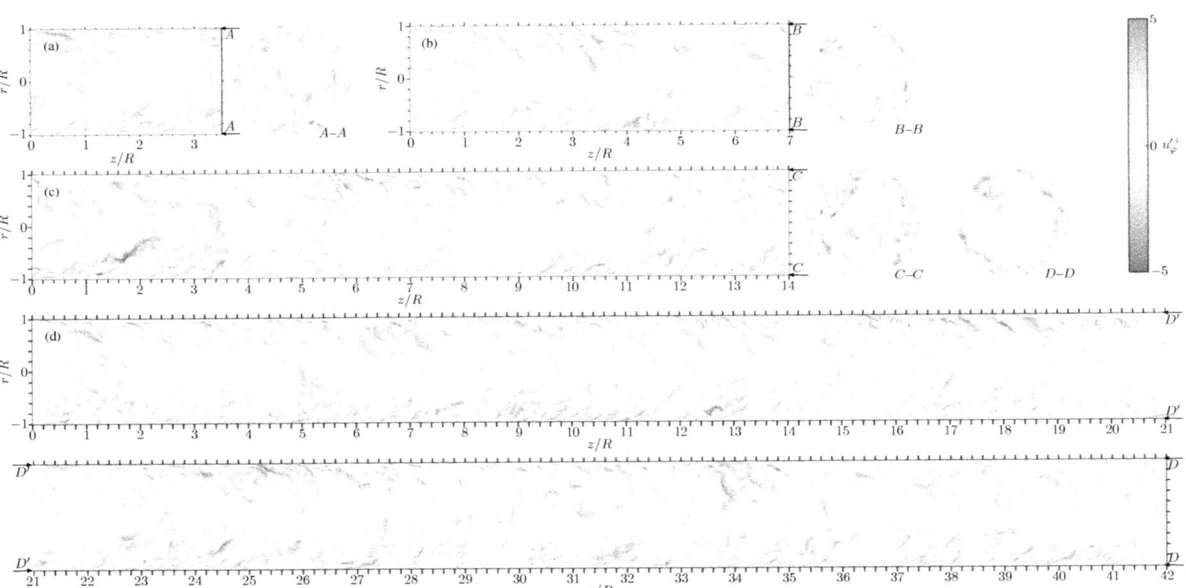

Figure 3.5: Instantaneous azimuthal fluctuating velocity u'^{+}_{φ} for different L in rz- and $r\varphi$-planes at $\mathrm{Re}_{\tau} = 720$. (a) P720L3; (b) P720L7 (C) P720L14; (d) P720.

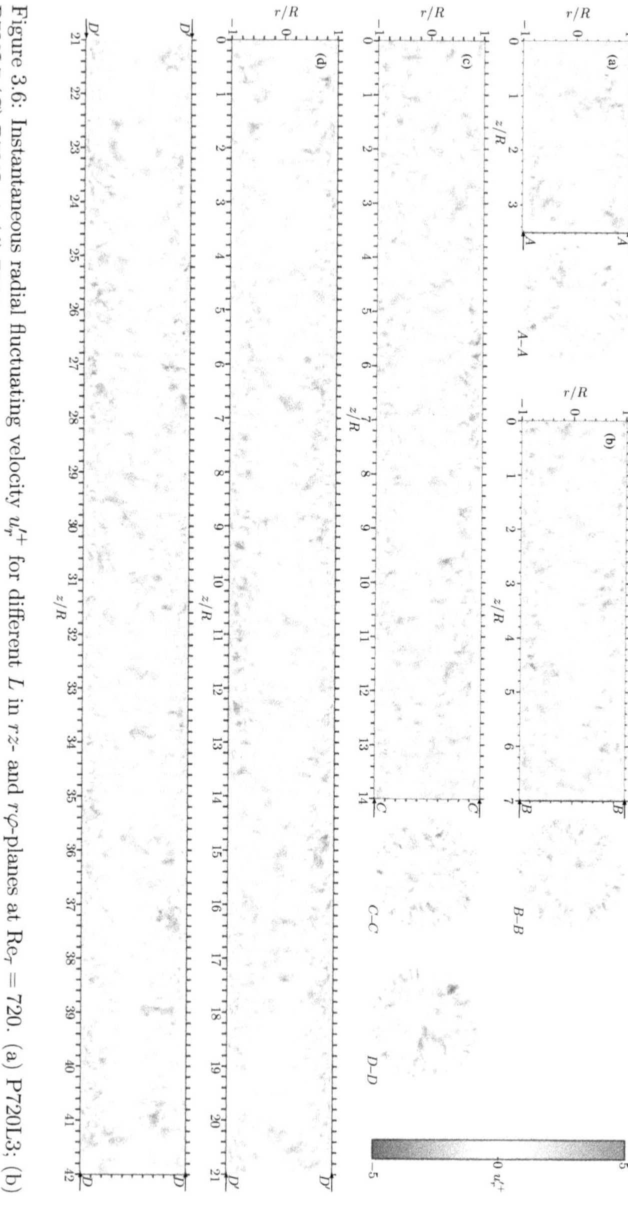

Figure 3.6: Instantaneous radial fluctuating velocity $u_r'^+$ for different L in rz- and $r\varphi$-planes at $\mathrm{Re}_\tau = 720$. (a) P720L3; (b) P720L7 (C) P720L14; (d) P720.

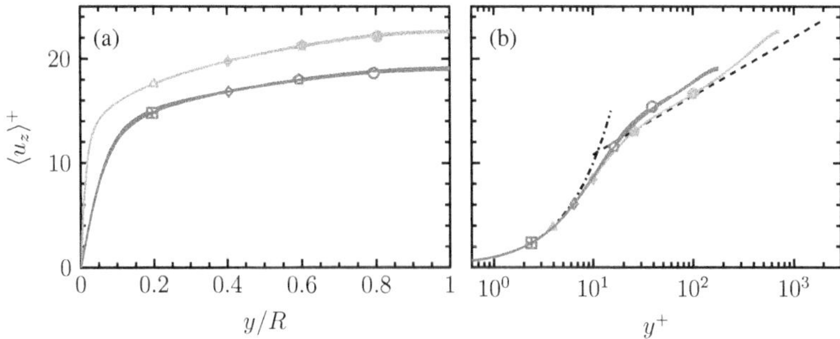

Figure 3.7: Mean streamwise velocity profile $\langle u_z \rangle^+$ for different L. (a), y-coordinate linear in bulk units; (b), y-coordinate logarithmic in wall units. -o-, P180L3; -o-, P180L7; -◇-, P180L14; -□-, P180; -+-, P180L84; -●-, P720L3; -●-, P720L7; -◆-, P720L14, -△-, P720. In (b): - - -, linear law of the wall $\langle u_z \rangle^+ = y^+$; ···, von Kármán log law: $\langle u_z \rangle^+ = 1/0.41 \log y^+ + 5.2$.

Regarding the mean velocity profile, depicted in figure 3.7 for $\mathrm{Re}_\tau = 180$ (blue lines) and $\mathrm{Re}_\tau = 720$ (green lines), only marginal variations with the domain length are detected. For $\mathrm{Re}_\tau = 720$, the centreline value of the mean streamwise velocity $\langle u_z \rangle|_{r=0} = 22.5$ lies slightly below the value of 22.6 obtained from longer numerical domains. For $\mathrm{Re}_\tau = 180$, the mean velocity profile obtained from the smallest domain length $L = 3.5R$ differs only marginally from the other profiles. While the latter overestimates the mean velocity in the buffer layer slightly ($\langle u_z \rangle|_{y^+=15} = 11.3$ compared to 11.1 for $L \geq 7R$), it underestimates the mean velocity at the pipe axis about the same amount ($\langle u_z \rangle|_{r=0} = 18.9$ compared to 19.1 for $L \geq 7R$). In summary, there is no influence of the domain length L on the mean velocity profile for $L \geq 7R$. Moreover, for $L = 3.5$ the domain length dependency of the mean velocity is only marginal.

Profiles of the turbulence intensities, shown in figure 3.8, are completely independent of the domain length above $L \geq 14R$ for $\mathrm{Re}_\tau = 180$ and $L \geq 7R$ for $\mathrm{Re}_\tau = 720$. For both Reynolds numbers the domain-length dependency is most pronounced for the streamwise profile, whereas the wall-normal profile shows only little ($\mathrm{Re}_\tau = 180$) or no dependency on the domain length ($\mathrm{Re}_\tau = 720$). Moreover, the largest deviation from longer domain curves (of about $0.25u_\tau$) is located in the buffer layer around $y^+ \approx 15$ for $\mathrm{Re}_\tau = 180$, whereas for $\mathrm{Re}_\tau = 720$ the profile differs most in the outer-flow region (approximately $0.08u_\tau$).

The domain-length dependency of high-order statistics is analysed via the profiles of the velocity skewness (figure 3.9) and the velocity flatness (figure 3.10). The former

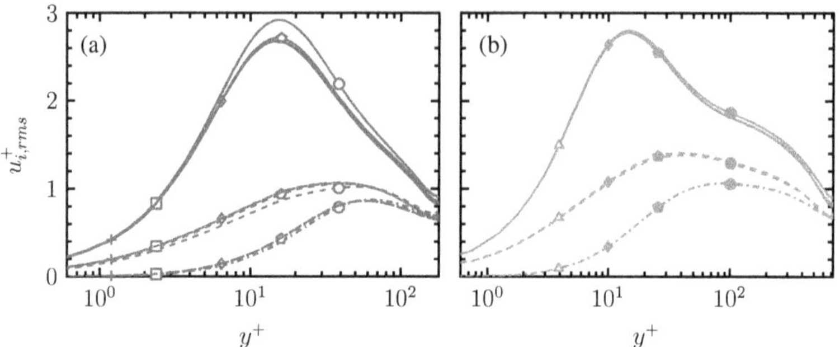

Figure 3.8: Turbulence intensities in wall units $u^+_{z,rms}$ (solid lines), $u^+_{\varphi,rms}$ (dashed lines), and $u^+_{r,rms}$ (dashed-dotted lines) for different L. (a), $\mathrm{Re}_\tau = 180$; (b), $\mathrm{Re}_\tau = 720$. -o-, P180L3; -o-, P180L7; -◇-, P180L14; -□-, P180; -+-, P180L84; -●-, P720L3; -●-, P720L7; -●-, P720L14, -△-, P720.

behave similar to the profiles of the turbulence intensities, which are independent of L for $L \geq 14R$ for $\mathrm{Re}_\tau = 180$ and $L \geq 7R$ for $\mathrm{Re}_\tau = 720$. As for the Reynolds stresses, the influence of the domain length is most pronounced in the streamwise component of the velocity skewness.

For the velocity flatness (figure 3.10), the influence of the domain length is most prominent for the wall-normal component at the wall. While the flatness profiles at $\mathrm{Re}_\tau = 720$ are already converged for $L \geq 7R$ — similar to lower order statistics —, the value of the wall-normal flatness profile in the vicinity of the wall at $\mathrm{Re}_\tau = 180$ required domain lengths of $L \geq 42R$ to converge. However, the latter effect is only restricted to a very thin layer at the wall ($y^+ \lesssim 5$). Above this layer, the flatness profiles at $\mathrm{Re}_\tau = 180$ are converged for $L \geq 7R$, as for $\mathrm{Re}_\tau = 720$. As will be discussed in section 4.2, the large wall-normal flatness value near the wall in turbulent pipe flow is not only sensitive to a change in the domain length L but also to a change in Reynolds number Re_τ and averaging interval ΔT. This behaviour is further elaborated in the upcoming chapters.

Since one-point statistics provide only integral information about the structure of the flow, two-point statistics are investigated in the following. Particularly, statistical footprints of VLSM and their interference with the computational domain are considered here. Therefore, velocity correlations and pre-multiplied energy spectra are computed and analysed. For $\mathrm{Re}_\tau = 180$, figure 3.11 displays two-point velocity correlations in the vicinity of the wall ($y^+ = 15$, figure 3.11a) and in the bulk flow region ($y/R = 0.4$,

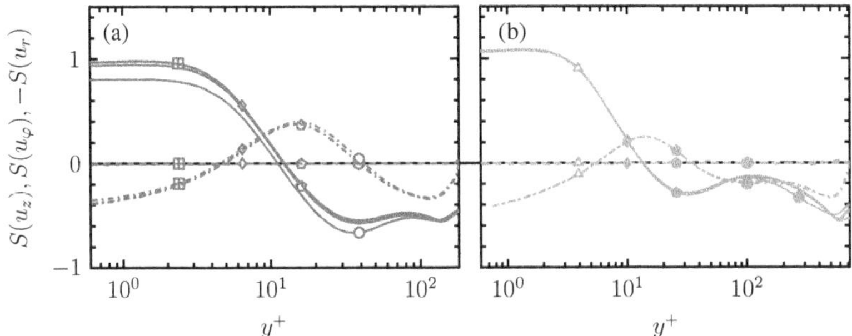

Figure 3.9: Profiles of the velocity skewness $S(u_z)$ (solid lines), $S(u_\varphi)$ (dashed lines), and $-S(u_r)$ (dashed-dotted lines) for for different L. (a), $\mathrm{Re}_\tau = 180$; (b), $\mathrm{Re}_\tau = 720$. -○-, P180L3; -⬠-, P180L7; -◇-, P180L14; -□-, P180; -+-, P180L84; -●-, P720L3; -⬟-, P720L7; -◆-, P720L14, -△-, P720.

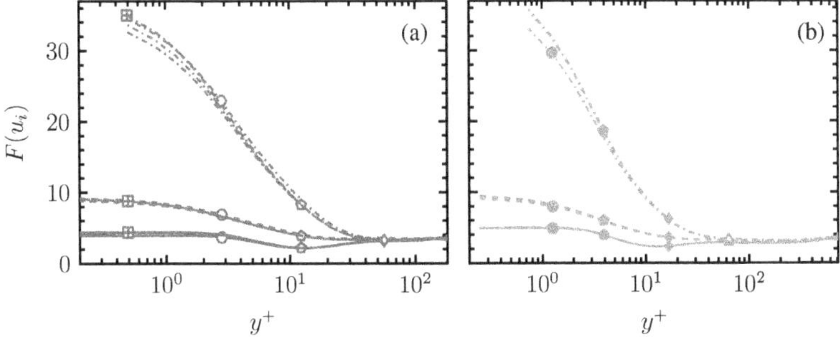

Figure 3.10: Profiles of the velocity flatness $F(u_z)$ (solid lines), $F(u_\varphi)$ (dashed lines), and $F(u_r)$ (dashed-dotted lines) for different L. (a), $\mathrm{Re}_\tau = 180$; (b), $\mathrm{Re}_\tau = 720$. -○-, P180L3; -⬠-, P180L7; -◇-, P180L14; -□-, P180; -+-, P180L84; -●-, P720L3; -⬟-, P720L7; -◆-, P720L14, -△-, P720.

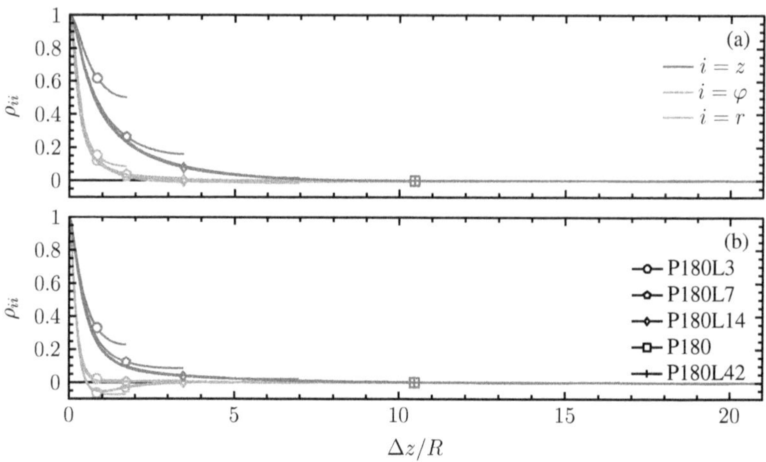

Figure 3.11: Two-point velocity correlations as a function of the streamwise separation length $\Delta z/R$ for different L at $\mathrm{Re}_\tau = 180$. (a) $y^+ = 15$; (b) $y/R = 0.4$.

figure 3.11b). In agreement with instantaneous observations (figures 3.1–3.3), the streamwise velocity correlation converges at larger computational domain lengths L than the correlation of the other velocity components. In particular, the streamwise velocity correlation is found to be independent of the numerical box size for $L \geq 14R$ both near the wall ($y^+ = 15$, figure 3.11a) and in the bulk region of the flow ($y/R = 0.4$, figure 3.11b). Near the wall, the profiles of the azimuthal velocity collapse for all L considered, whereas the wall-normal profiles are independent of the domain length for $L \geq 7R$ (figure 3.11a). Away from the wall, on the contrary, all wall-normal profiles collapse, whereas the azimuthal profiles are converged above $L \geq 7R$ figure 3.11b. Above all, the streamwise velocity correlation is most sensitive to a change of the domain length L, because the underlying energetic coherent structures are particularly elongated in streamwise direction. Figure 3.12 shows iso-surfaces of the three-dimensional two-point velocity correlation of the streamwise velocity component for different L at $y^+ = 15$, which reflect the three-dimensional shape of the streamwise elongated coherent structures. As visible in figure 3.12(a,b), these structures interact with the computational box, if the latter is chosen to be insufficiently large. To compare the shape of the streamwise elongated structures more quantitatively, iso-contours of two-dimensional streamwise velocity correlation in the rz-plane are

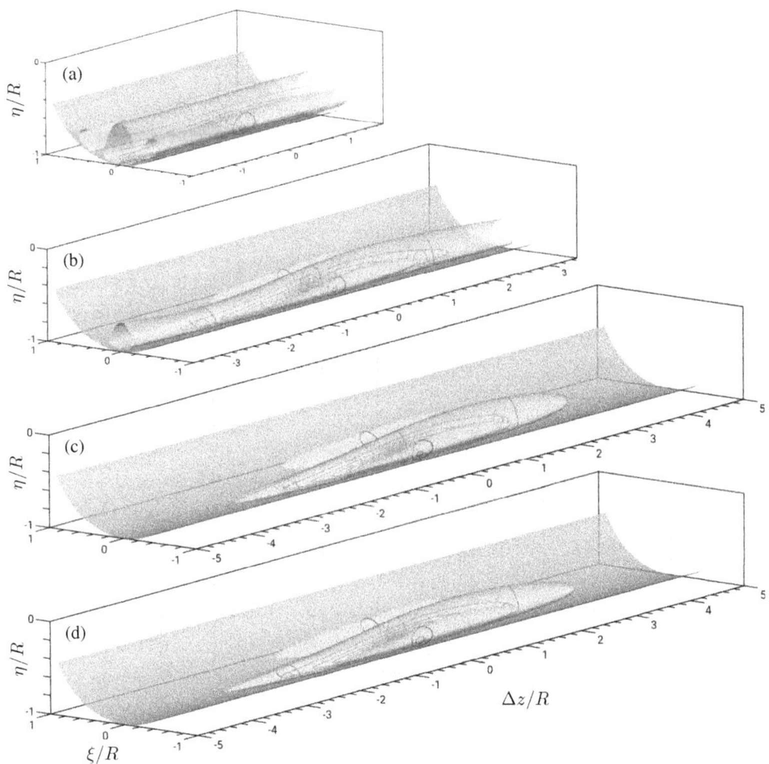

Figure 3.12: Iso-surfaces and iso-contours of the three-dimensional two-point velocity correlation of the streamwise velocity component $\rho_{zz}(\Delta z, \Delta\varphi, \Delta r, r_0)$ as a function of the axial, azimuthal and radial separation lengths for different L. Reference point of the correlation at $y_0^+ = 15$. Cartesian cross-sectional coordinates $\xi = (r_0 + \Delta r)\sin(\Delta\varphi)$, $\eta = (r_0 + \Delta r)\cos(\Delta\varphi)$. (a) P180L3; (b) P180L7; (c) P180L14; (d) P180. The orange (cyan) iso-surfaces exhibit values of $+(-)0.1$. Iso-contours values range from $+(-)0.1$ to $+(-)1$, increment of 0.1.

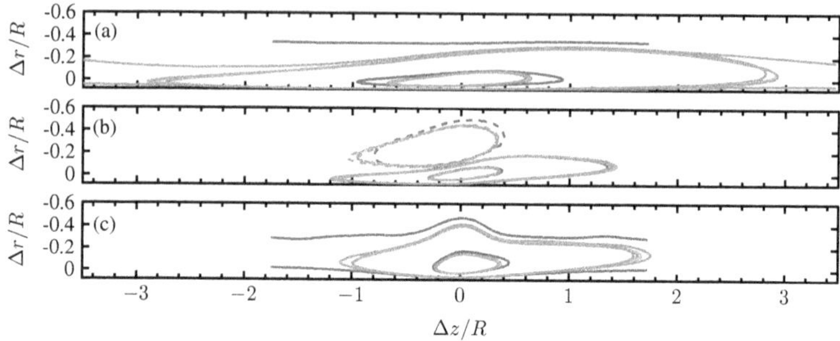

Figure 3.13: Iso-contours of the two-dimensional two-point velocity correlation $\rho_{ii}(\Delta z, \Delta r)$ as a function of the axial and radial separation lengths for different L. Reference point of the correlation at $y_0^+ = 15$. —, P180L3; —, P180L7; —, P180L14; —, P180. (a) ρ_{zz}; (b) $\rho_{\varphi\varphi}$; (c) ρ_{rr}. Solid (dashed) iso-contours indicate values of $+(-)0.62$ and $+(-)0.12$. Cartesian cross-sectional coordinates $\xi = (r_0 + \Delta r)\sin(\Delta\varphi)$, $\eta = (r_0 + \Delta r)\cos(\Delta\varphi)$.

presented in figure 3.13(a) together with correlation iso-contours of the other velocity components (figures of three-dimensional correlation iso-surfaces of the latter being omitted). Figure 3.13(a) reveals that the characteristic structure inclination is recovered to a certain extent for $L = 7R$, while there is no inclination visible for $L = 3.5R$, both correlations not decaying to zero. The negative correlated region at $(\Delta z/R = 0, \Delta r/R < -0.2)$ in the azimuthal velocity correlation, displayed in figure 3.13(b), suggest that azimuthal velocity structures are only converged for $L \geq 7R$, which is not visible when considering the one-dimensional correlation in streamwise direction at $y^+ = 15$, only (see figure 3.11a). In agreement with its one-dimensional representation in streamwise direction (figure 3.11a), the radial velocity structure in the rz-plane in figure 3.13(c) is independent of the computational domain length for $L \geq 7R$. The radial velocity component appears to be more affected by a very short computational domain ($L = 3.5R$) than the azimuthal component at $\mathrm{Re}_\tau = 180$ and $y_0^+ = 15$. The cross-sectional shape of the streamwise coherent structures for different L is compared via iso-contours of the two-dimensional streamwise velocity correlation in a cross-sectional plane displayed in figure 3.14. For the shortest computational domain ($L/R = 3.5$), the streamwise velocity component structure becomes thicker in azimuthal and radial direction and the azimuthal spacing between adjacent structures increases when compared with structures from larger computational boxes (figure 3.14). For $L/R = 7$ the streamwise velocity component structure is only slightly higher than

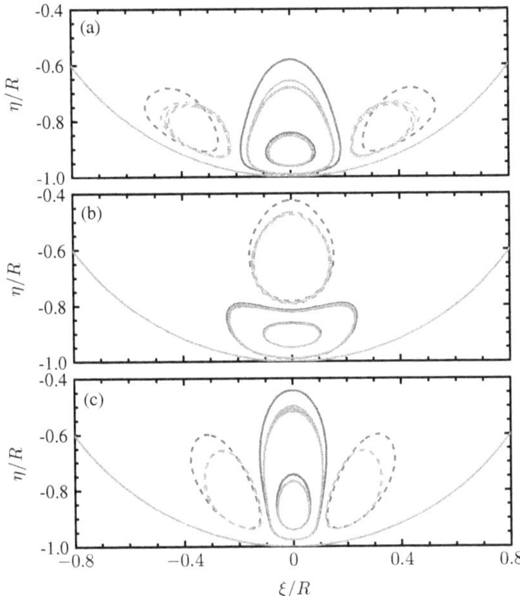

Figure 3.14: Iso-contours of the two-dimensional two-point velocity correlation $\rho_{ii}(\Delta z, \Delta\varphi)$ as a function of the azimuthal and radial separation lengths for different L. Reference point of the correlation at $y_0^+ = 15$. —, P180L3; —, P180L7; —, P180L14; —, P180. (a) ρ_{zz}; (b) $\rho_{\varphi\varphi}$; (c) ρ_{rr}. Solid (dashed) iso-contours indicate values of $+(-)0.62$ and $+(-)0.12$. Cartesian cross-sectional coordinates $\xi = (r_0 + \Delta r)\sin(\Delta\varphi)$, $\eta = (r_0 + \Delta r)\cos(\Delta\varphi)$.

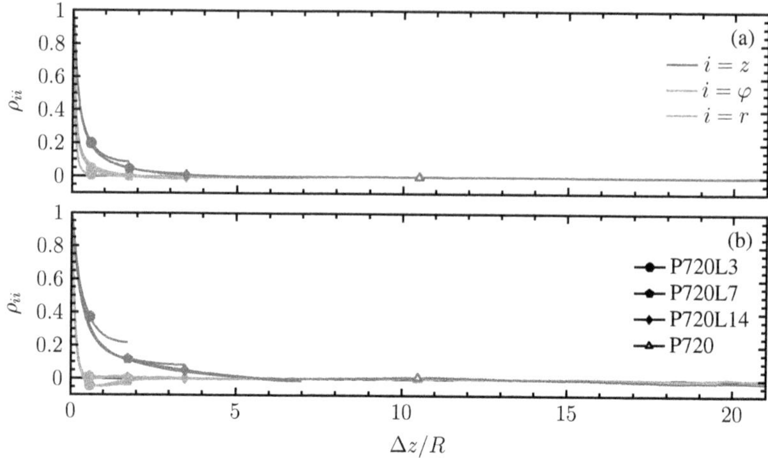

Figure 3.15: Two-point velocity correlations as a function of the streamwise separation length $\Delta z/R$ for different L at $\mathrm{Re}_\tau = 720$. (a) $y^+ = 15$; (b) $y/R = 0.4$.

structures from $L/R > 7$. While the cross-sectional correlation iso-contours of the streamwise velocity component are independent of the domain length for $L/R \geq 14$ (figure 3.14), the iso-contours of the azimuthal and radial velocity component are already converged at $L/R \geq 7$ (figure 3.14b,c). The effect of the short computational domain length ($L/R = 3.5$) appears to be larger on the radial velocity component structure (figure 3.14c) than on the azimuthal velocity component structure (figure 3.14b).

For $\mathrm{Re}_\tau = 720$, one-dimensional two-point velocity correlations are depicted in figure 3.15. While the streamwise velocity correlation in the buffer layer is already converged for $L \geq 7R$, the azimuthal and wall-normal correlations at the same wall distance collapse for all domain lengths considered (figure 3.15a). As indicated by figure 3.15(b), in the outer-flow region at $y/R = 0.4$, domains of $L \geq 14R$ are required to obtain a converged streamwise velocity correlation. Iso-surfaces of the three-dimensional streamwise velocity correlation with a reference point in the bulk flow region ($y_0/R = 0.4$) are displayed in figure 3.16. In addition to the one-dimensional streamwise velocity correlation (figure 3.15), figure 3.16 shows that negative correlated regions (blue iso-surfaces) are elongated in streamwise direction for $L/R = 3.5$ and shortened for $L/R = 7$. A quantitative comparison of velocity correlation iso-contours in an zr-plane at $\Delta\varphi = 0$ is presented in figure 3.17. The iso-contours of the streamwise velocity component

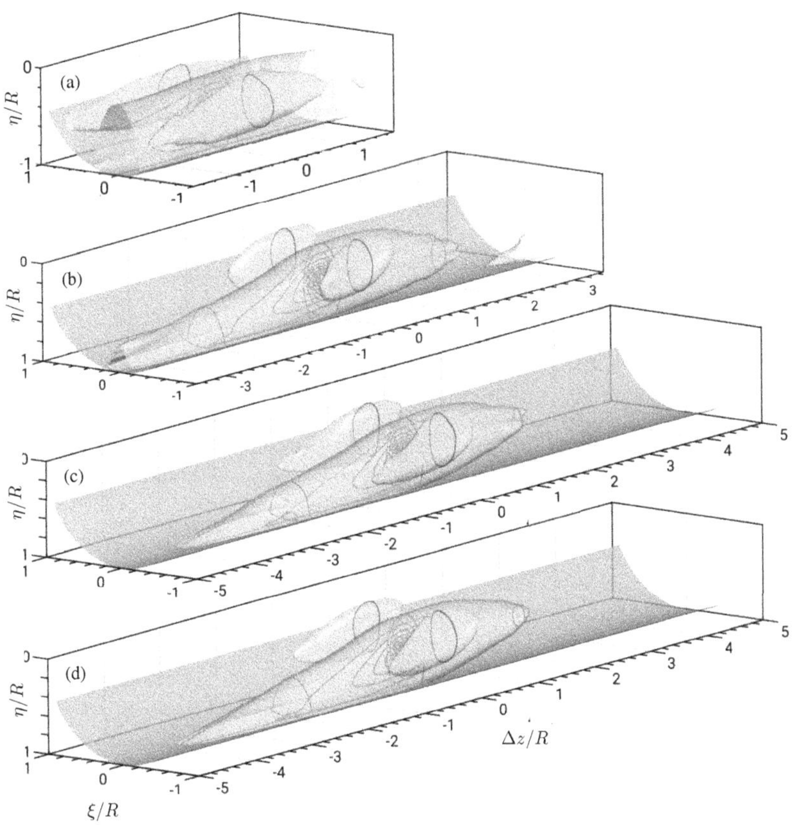

Figure 3.16: Iso-surfaces and iso-contours of the three-dimensional two-point velocity correlation of the streamwise velocity component $\rho_{zz}(\Delta z, \Delta\varphi, \Delta r, r_0)$ as a function of the axial, azimuthal and radial separation lengths for different L. Reference point of the correlation at $y_0/R = 0.4$. Cartesian cross-sectional coordinates $\xi = (r_0 + \Delta r)\sin(\Delta\varphi)$, $\eta = (r_0 + \Delta r)\cos(\Delta\varphi)$. (a) P720L3; (b) P720L7; (c) P720L14; (d) P720. The orange (cyan) iso-surfaces exhibit values of $+(-)0.1$. Iso-contours values range from $+(-)0.1$ to $+(-)1$, increment of 0.1.

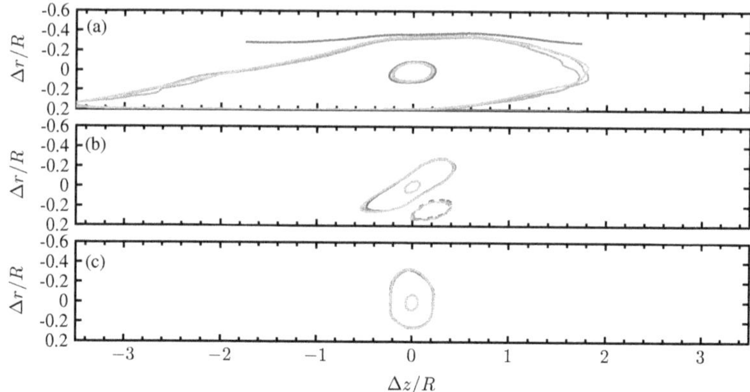

Figure 3.17: Iso-contours of the two-dimensional two-point velocity correlation $\rho_{ii}(\Delta z, \Delta r)$ as a function of the axial and radial separation lengths for different L. Reference point of the correlation at $y_0/R = 0.4$. —, P720L3; —, P720L7; —, P720L14; —, P720. (a) ρ_{zz}; (b) $\rho_{\varphi\varphi}$; (c) ρ_{rr}. Solid (dashed) iso-contours indicate values of $+(-)0.62$ and $+(-)0.12$. Cartesian cross-sectional coordinates $\xi = (r_0 + \Delta r)\sin(\Delta\varphi)$, $\eta = (r_0 + \Delta r)\cos(\Delta\varphi)$.

are fairly converged for $L/R \geq 7R$ (figure 3.17), wheres the iso-contours of the other two velocity components are already converged for the shortest computational domain length considered ($L/R = 3.5$, figure 3.17b,c). In the cross-sectional plane, depicted in figure 3.18, velocity correlation iso-contours indicate that the streamwise velocity component structures in the bulk flow region for $L/R = 3.5$ are higher and wider than the structures obtained from numerical boxes of size $L/R \geq 7R$ (figure 3.17a). Structures corresponding to the azimuthal and radial velocity component appear to be already converged for the smallest computational domain length $L/R = 3.5$ (figure 3.18b,c). Although velocity correlations reflect the structure of the flow, they are integral quantities in terms of contributions from different scales.

To further evaluate the statistically relevant interaction between turbulent coherent structures and the computational domain, two-dimensional pre-multiplied streamwise velocity spectra are presented in figure 3.19 at the two particular wall distances used above. In the buffer layer (figure 3.19a,c), the spectral peak at $\lambda_z^+ \approx 1000$, $\lambda_\varphi^+ \approx 120$ represents the velocity streaks. Since this species of coherent structure scales in viscous units, a bigger part of them is cut off by an insufficiently small computational domain for $\mathrm{Re}_\tau = 180$ (figure 3.19a) than for $\mathrm{Re}_\tau = 720$ (figure 3.19c). Thus, streamwise velocity

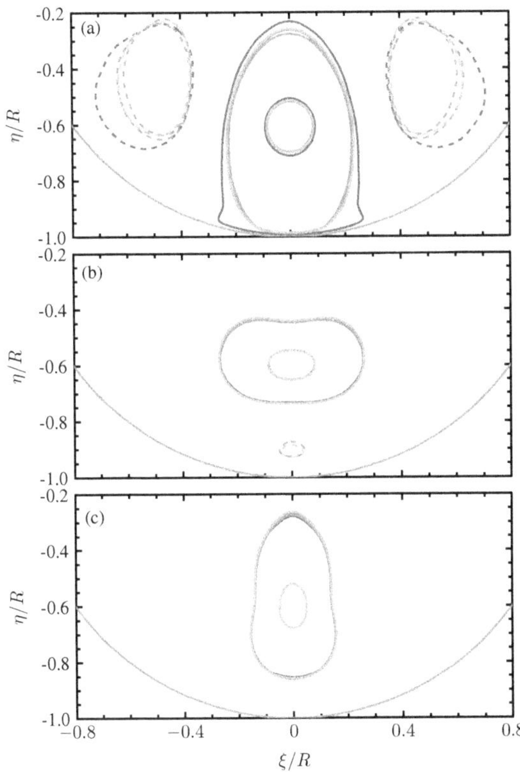

Figure 3.18: Iso-contours of the two-dimensional two-point velocity correlation $\rho_{ii}(\Delta z, \Delta\varphi)$ as a function of the azimuthal and radial separation lengths for different L. Reference point of the correlation at $y_0/R = 0.4$. —, P720L3; —, P720L7; —, P720L14; —, P720. (a) ρ_{zz}; (b) $\rho_{\varphi\varphi}$; (c) ρ_{rr}. Solid (dashed) iso-contours indicate values of +(−)0.62 and +(−)0.12. Cartesian cross-sectional coordinates $\xi = (r_0 + \Delta r)\sin(\Delta\varphi)$, $\eta = (r_0 + \Delta r)\cos(\Delta\varphi)$.

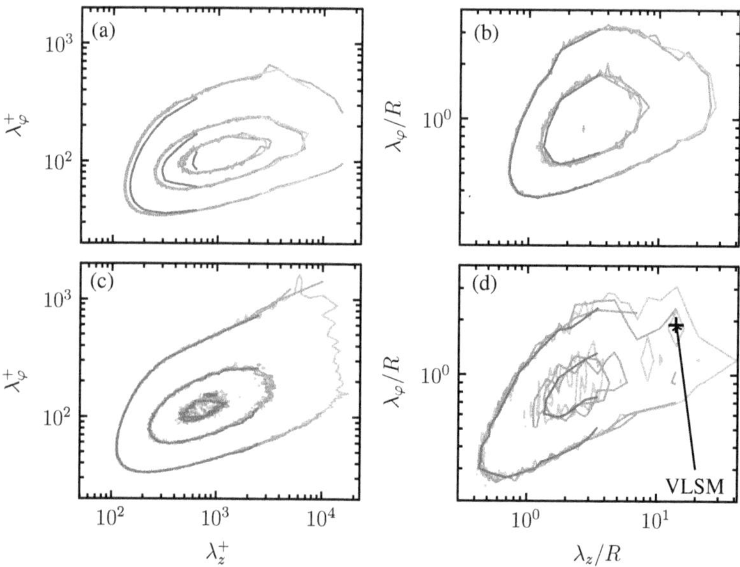

Figure 3.19: Two-dimensional pre-multiplied energy spectra of the streamwise velocity component $\kappa_z\kappa_\varphi\phi_{zz}$ as functions of the streamwise wavelength λ_z and the spanwise wavelength λ_φ for different L. All quantities normalised in wall units. (a,b), $\mathrm{Re}_\tau = 180$; —, P180L3; —, P180L7; —, P180L14; —, P180; —, P180L42. (c,d), $\mathrm{Re}_\tau = 720$; —, P720L3; —, P720L7; —, P720L14; —, P720. (a,c) $y^+ = 15$, iso-contour values range from 0.1 to 1.0 with an increment of 0.3; (b,d) $y \approx 0.4R$, iso-contours indicate values of 0.12 and 0.24.

correlations are affected up to larger domain sizes for $\text{Re}_\tau = 180$ in the vicinity of the wall. Away from the wall, the spectra for $\text{Re}_\tau = 180$ (figure 3.19b) still exhibits the footprint of velocity streaks, which cover a large fraction of the flow domain at such low Re_τ. For $\text{Re}_\tau = 720$ (figure 3.19d), additional coherent structures, i.e. LSM and VLSM, come into play. The spectral peak around $\lambda_z/R \approx 3$ and $\lambda_\varphi/R \approx 1$ in figure 3.19(d) represents LSMs, whereas the one at $\lambda_z/R \approx 15$ and $\lambda_\varphi/R \approx 2$ is the footprint of VLSMs[10]. Since the former type of coherent motion contributes more to the overall TKE than the latter at the Reynolds number considered here, velocity correlations are already fairly converged for $L/R \geq 14$ at $\text{Re}_\tau = 720$, albeit not capturing the spectral VLSM peak. Nonetheless, as the spectra only decay for large streamwise wavelengths ($\lambda_z/R \approx 42$, figure 3.19a) and since a particular aim of the following studies is the investigation of VLSM not only statistically but also in terms of instantaneous flow field realisations, these studies are set up in domains of $L/R = 42$.

Overall, in good agreement with the study of Chin et al. (2010), the mean velocity profile converges faster with the domain length L than higher order moments or two-point velocity correlations. Moreover is the domain length needed for convergence of higher order statistics Reynolds number dependent. Turbulence intensities and skewness factors are already converged at $L/R = 7$ for $\text{Re}_\tau = 720$ ($L/R = 2\pi$ for $\text{Re}_\tau = 500$ Chin et al. (2010)), whereas for $\text{Re}_\tau = 180$ it takes $L/R = 14$ ($L/R = 4\pi$ for $\text{Re}_\tau = 170$ Chin et al. (2010)). In addition, the wall-normal location, where the profiles exhibit differences shifts away from the wall for the larger Reynolds number case, which is presumably caused by the appearance of LSM and VLSM. Although the latter do not carry a substantial fraction of TKE for the Reynolds number considered, their footprint is visible in the streamwise pre-multiplied energy spectrum at wavelengths up to $\lambda_z/R = 42$. As Hoyas and Jiménez (2006) and del Álamo et al. (2004) suggested for turbulent plane channel flow, Chin et al. (2010) proposed a pipe length of $8\pi R$ necessary for turbulence statistics to converge up to $\text{Re}_\tau = 500$. For even larger Reynolds numbers, where VLSM become increasingly energetic, the current study suggest domain length of $L/R = 42$.

[10]Note that for higher Reynolds numbers, the large streamwise wavelength peak is shifted to even higher values (shown in figure 4.30), where it presumably saturates.

Chapter 4

Scaling and Interaction of Coherent Structures in Turbulent Pipe Flow

In the upcoming chapter, the scaling of turbulence in pipe flow is investigated by means of turbulence statistics and underlying coherent structures. Particular attention is payed to high-order velocity statistics in the vicinity of the wall. First, the coherent structure of the flow is described via instantaneous flow field realisations (section 4.1). Subsequently, the convergence and scaling of turbulence one-point statistics is analysed (section 4.2), before the scaling of two-point statistical quantities is elaborated in section 4.3. Finally, the interactions between different coherent structures are analysed with the aid of spatial filtering (section 4.4) and conditional averaging (section 4.5). Major parts of the analysis have been published in Bauer et al. (2017, 2018), and Bauer and Wagner (2019b). As presented in table 4.1, the analysis involves five different $Re_\tau \in \{180, 360, 720, 1500, 2880\}$ and three different grid refinement levels for the lowest Reynolds number.

4.1 Instantaneous Flow Field Observation

In the following, the flow structure is analysed via time series and instantaneous flow field realisations. First, the phenomenology of velocity spikes is discussed, before the coherent structure of the streamwise velocity component is elaborated.

4.1.1 Velocity Spikes

As already mentioned in section 1.2.5, Xu et al. (1996) showed that very strong wall-normal velocity fluctuations, which occur seldom in space and time, contribute most

Table 4.1: Turbulent pipe-flow simulation cases. N_z, N_φ and N_r are the number of grid points with respect to the axial, azimuthal and radial direction, respectively. Δz^+, streamwise grid spacing; $R^+\Delta\varphi$, azimuthal grid spacing at the wall; Δr^+_{min} and Δr^+_{max}, minimal and maximal radial grid spacing, respectively, all grid spacings normalised by wall units. ΔT, averaging interval for statistics; $\Delta T_b = \Delta T u_b/R$, in bulk time units; $\Delta T^+ = \Delta T u_\tau^2/\nu$, in viscous time units; $\Delta\tau^+ = \Delta T^+ L^+ 2\pi r^+$, averaging coordinate in viscous units.

Case	P180C	P180	P180F	P360	P720	P1500	P2880
line	- - -	-□-	···	-○-	-△-	-◇-	-⊗-
Re_τ	180	180	180	360	720	1500	2880
Re_b	5258	5258	5258	11 664	25 992	60 075	123 260
L/R	42	42	42	42	42	42	42
N_z	1536	1536	1536	3072	4608	8192	12 288
N_φ	256	256	256	512	1024	2048	4096
N_r	76	84	104	160	222	408	646
Δz^+	4.9	4.9	4.9	4.9	6.6	7.7	9.8
$R^+\Delta\varphi$	4.4	4.4	4.4	4.4	4.4	4.6	4.4
Δr^+_{min}	0.50	0.31	0.11	0.39	0.49	0.49	0.32
Δr^+_{max}	4.4	4.4	4.4	4.4	6.6	7.8	4.4
ΔT_b	9826	29 396	3641	2838	788	310	27
ΔT^+	121 050	361 880	44 880	63 074	31 450	23 230	3650
$\Delta\tau^+ \cdot 10^{12}$	1.04	3.09	0.38	2.16	4.30	13.8	8.0

to the large wall-normal flatness value near the wall. Figure 4.1 shows a time series of the wall-normal velocity at $y^+ = 2$ obtained from P180, which includes three events of positive and negative velocity spikes within 1600 viscous time units. Focussing on the positive velocity spike around $t^+ \approx 908$, a segment of the same time series — here normalised by the local RMS value — is shown in figure 4.2(a) for wall distances of $y^+ = 2$ and $y^+ = 49$ together with series of the streamwise velocity fluctuation u'_z (figure 4.2b) and the Reynolds shear stress $u'_z u'_r$ (figure 4.2c). While the positive velocity spike at $t^+ \approx 908$ is pronounced in the time series close to the wall, no such event is visible in the time series at $y^+ = 49$ (figure 4.2a). Moreover, the time series of the streamwise velocity fluctuation exhibits a strong positive fluctuation at $t^+ \approx 908$ and $y^+ = 2$ (figure 4.2b), which leads to a strong turbulent sweep in the time series of the Reynolds shear stress at $t^+ \approx 908$ and $y^+ = 2$ (figure 4.2c). The time series at $y^+ = 49$ in figure 4.2 do not exhibit extreme events as the ones at $y^+ = 2$ close to wall indicating that the phenomenology of velocity spikes is limited close to the wall. The spatial environment around the velocity spike at $t^+ \approx 908$ (figure 4.2a) is presented in figure 4.3, where three orthogonal planes cutting the probe location at $y^+ = 2$ are

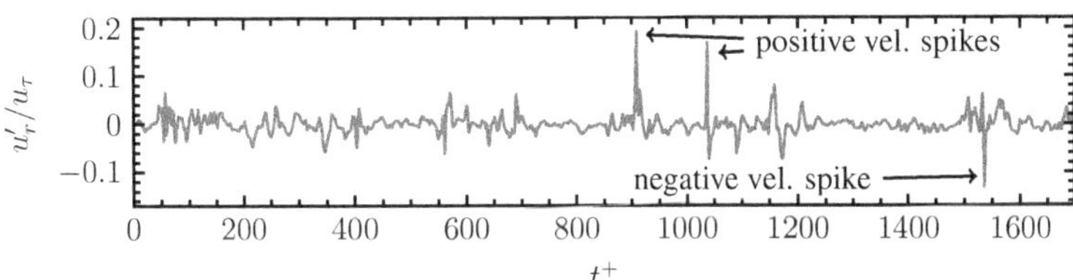

Figure 4.1: Time series of u'_r, normalised by u_τ, at a wall-normal positions of $y^+ = 2$ for P180.

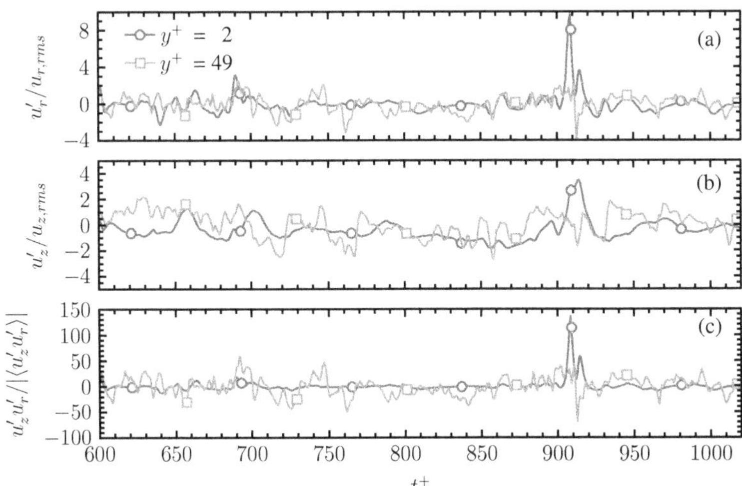

Figure 4.2: Time series of u'_r (a), u'_z (b), $u'_z u'_r$ (c), normalised by the corresponding RMS value recorded at wall-normal positions $y^+ = 2$ and 49 for P180.

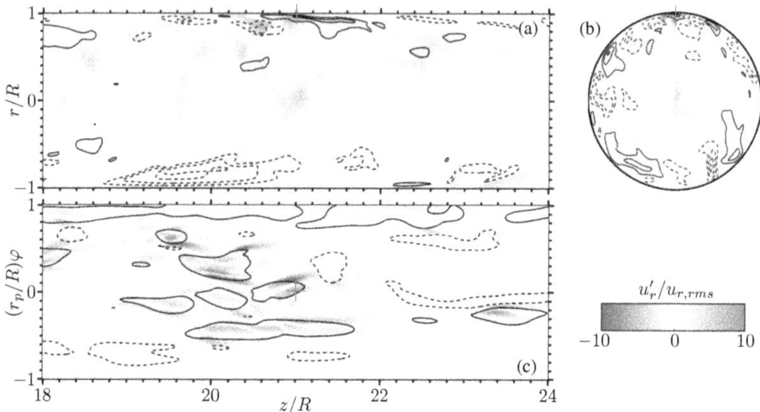

Figure 4.3: Instantaneous snapshots of the wall-normal velocity component at $t^+ = 905$ in planes cutting the probe location (+) of the time series shown in figure 4.2 for P180. Black lines show iso-contours of positive (—) and negative (- - -) streamwise velocity fluctuations u'_z normalised by u_τ. (a,b) Iso-contours have an increment of 2 ranging from −6 to 6. (c) Increment of 0.8, ranging from −0.8 to 0.8.

shown. As the $z\varphi$-plane in figure 4.3(b) indicates, positive velocity spikes (dark red) predominately coincide with regions of positive streamwise fluctuations (solid lines). The latter appear to be footprints of high-speed velocity streaks at distances further away from the wall, shown as solid lines in the rz-plane in figure 4.3(a). Negative velocity spikes (dark blue), on the contrary, are rather located in between regions of high and low streamwise momentum. Moreover, figure 4.3(c) shows that the velocity spikes in the region $-0.6 \leq (r_p/R)\varphi \leq 0.7$, $19 \leq z/R \leq 22$ cluster in pairs of positive and negative spikes. In agreement with the time series depicted in figure 4.2, figure 4.3(a,b) reveals that velocity spikes are settled in the vicinity of the wall.

4.1.2 Velocity Streaks and VLSMs

The interaction of large-scale outer-flow motions with the near-wall region, first introduced in section 1.2.5, reflects in instantaneous streamwise velocity fluctuations as presented in figures 4.4 and 4.5 and is elaborated in the following. First, figure 4.4 shows the streamwise velocity fluctuation in a wall-parallel plane at $y^+ = 15$ for the different Reynolds numbers considered. For $\mathrm{Re}_\tau = 180$, high- and low-speed streaks are homogeneously distributed within the wall-parallel plane (figure 4.4a). When increasing

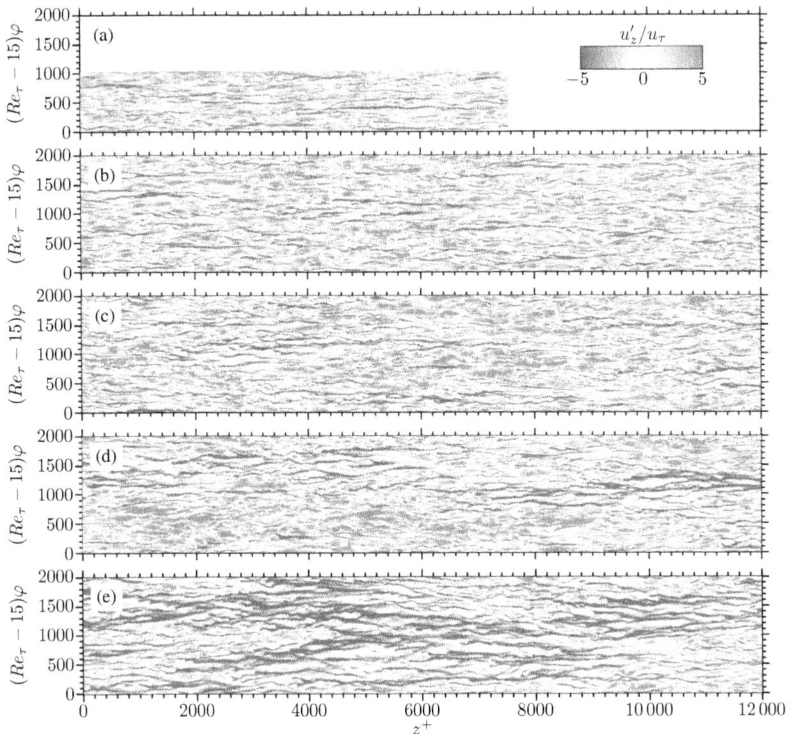

Figure 4.4: Iso-contours of streamwise velocity fluctuations in a wall-parallel pipe segment of size $l_z^+ \times l_\varphi^+ = 12000 \times 2000$ at $y^+ = 15$. (a), $\mathrm{Re}_\tau = 180$; (b), $\mathrm{Re}_\tau = 360$; (c), $\mathrm{Re}_\tau = 720$; (d) $\mathrm{Re}_\tau = 1500$. (e) $\mathrm{Re}_\tau = 2880$.

the Reynolds number, the velocity streaks organise in clusters, as visible in figure 4.4(b,c,d,e) for $\mathrm{Re}_\tau \geq 360$. For $\mathrm{Re}_\tau = 360$ only a small cluster of low-speed structures appears around $(\mathrm{Re}_\tau - 15)\varphi \approx 1000$, $z^+ \approx 4000$ (figure 4.4b), whereas for $\mathrm{Re}_\tau = 360$ a larger low-speed cluster is detected around $(\mathrm{Re}_\tau - 15)\varphi \approx 1000$, $z^+ \approx 3000$ (figure 4.4c). Finally, the upper half of the segment depicted for $\mathrm{Re}_\tau = 1500$ and the full segment depicted for $\mathrm{Re}_\tau = 2880$ are covered by low-speed clusters (figure 4.4d,e). Next, the streamwise velocity structures in the buffer layer are compared with structures of the same flow field realisation at a wall distance $y/R = 0.2$ presented in figure 4.5. At this particular distance from the wall, in the upper part logarithmic layer, the streaky structures have grown in size with respect to their counterparts in the buffer layer (figure 4.4). For $\mathrm{Re}_\tau = 180$, however, these LSMs are of comparable size to the SSMs in the buffer layer (figure 4.5a). With increasing Re_τ, the structures at $y/R = 0.2$ increase in length and width when normalised in wall units as the range of scales increases (figure 4.5). For $\mathrm{Re}_\tau \geq 1500$, very long intense streamwise velocity fluctuations manifest themselves. The interaction of these outer-flow motions with the near-wall region reflects in the clustering of wall-layer streaks mentioned above. For $\mathrm{Re}_\tau = 1500$ the footprint of the outer-flow low-speed structure (figure 4.5d) is visible as the low-speed streak cluster in the vicinity of the wall (figure 4.4d). To measure the full spatial extent of the low-speed structure in figure 4.5(d), this particular structure is extracted from the flow field and shown in figure 4.6 as a three-dimensional iso-surface. As indicated in figure 4.6, the single extracted low-speed flow structures measures approximately $l = 22R$ in streamwise length and is, thus, a representative of the species of VLSMs. The picture is complemented by the global image of streamwise velocity fluctuation iso-volumes obtained from the same flow field realisation, depicted in an interior view of the pipe geometry in figure 4.7. The iso-volumes represent again structures of positive and negative velocity fluctuations. Three azimuthally alternating high- and low-speed VLSMs clearly dominate the scenario. Since instantaneous observations do not necessarily represent the general structure of the flow, turbulence statistics are considered hereinafter. First, one-point statistics are investigated in section 4.2 before the average shape of the above-described coherent structures is determined by means of two-point statistics in section 4.2. Finally, the interactions between coherent structures are elaborated in section 4.4 and 4.5.

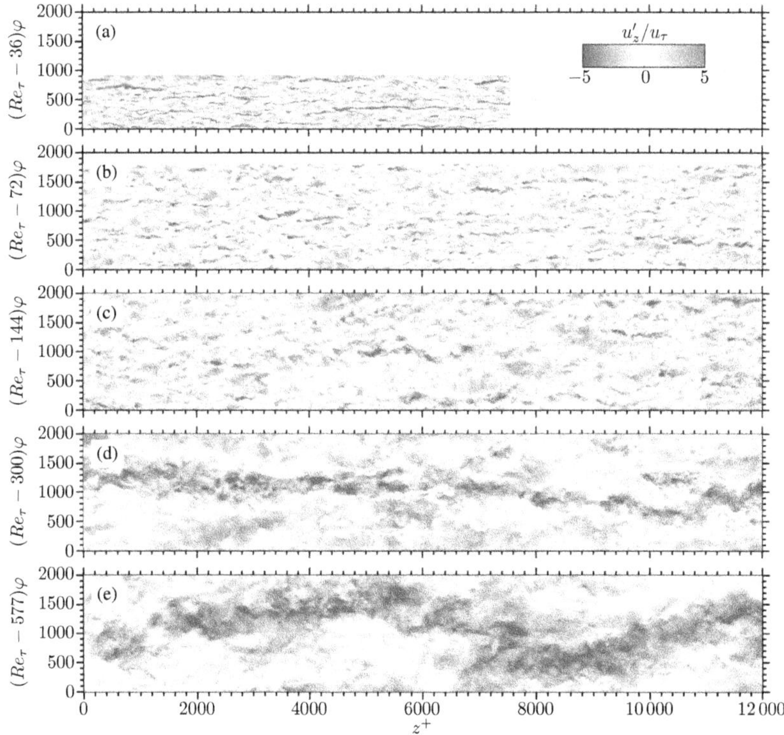

Figure 4.5: Iso-contours of streamwise velocity fluctuations in a wall-parallel pipe segment of size $l_z^+ \times l_\varphi^+ = 12000 \times 2000$ at $y/R = 0.2$. (a), $\mathrm{Re}_\tau = 180$; (b), $\mathrm{Re}_\tau = 360$; (c), $\mathrm{Re}_\tau = 720$; (d) $\mathrm{Re}_\tau = 1500$; (e) $\mathrm{Re}_\tau = 2880$.

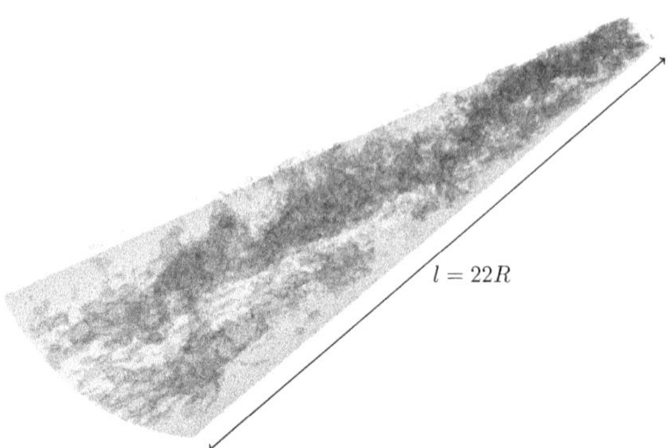

Figure 4.6: Low speed VLSM extracted from the instantaneous snapshot also visible in figure 4.4(d) and 4.5(d). Red (blue) semi-transparent iso-surfaces show streamwise velocity fluctuations $u_z'^+ = +(-)3$. $\mathrm{Re}_\tau = 1500$.

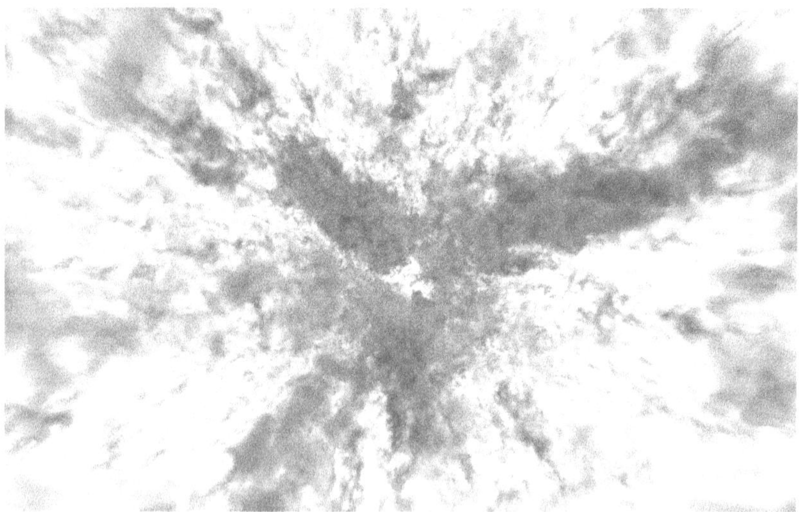

Figure 4.7: Iso-volumes of the streamwise velocity fluctuation $u_z'^+$. Red (blue) iso-volumes exhibit values ranging from $+(-)3$ to $+(-)5$. $\mathrm{Re}_\tau = 1500$.

4.2 One-point Statistics

The convergence and scaling of one-point statistics in turbulent pipe flow for the simulation cases presented in table 4.1 is investigated in the following. Recalling expression (1.33), a statistical average is obtained by integrating in time as well as over both homogeneous directions. Since the current analysis focuses on the convergence and scaling of near-wall statistics, averaging intervals both in space and time are normalised by the viscous length scale and time scale, respectively. Hereinafter, I assume that the convergence of near-wall statistics scales with the averaging coordinate

$$\Delta\tau^+ = \Delta T^+ L^+ 2\pi r^+, \tag{4.1}$$

independent of Reynolds number and/or domain length. Here $\Delta T^+ = \Delta T u_\tau^2/\nu$ is the time interval and L^+ as well as $2\pi r^+$ are the spatial intervals, all of whom being normalised in wall units.

4.2.1 Convergence of One-point Statistics

In the first step, both the statistical convergence and the grid convergence of turbulence statistics are investigated for turbulent pipe flow at $\mathrm{Re}_\tau = 180$. The latter involves three different wall-normal grid refinement levels of $\Delta r^+_{min} = 0.5$ (P180C), $\Delta r^+_{min} = 0.31$ (P180), and $\Delta r^+_{min} = 0.11$ (P180F). For the investigation of the statistical convergence, on the other hand, the temporal averaging interval is varied, while spatial averaging intervals are kept constant in bulk units.

Figure 4.8 shows PDFs of the three velocity components for P180 at six different distances from the wall. These PDFs reflect the non-Gaussian behaviour of wall-bounded turbulence. First, the PDFs of the streamwise fluctuating velocity are skewed to right for $y^+ \leq 10$ and to the left for $y^+ \geq 20$ (figure 4.8a). The PDFs of the azimuthal velocity presented in figure 4.8(b), on the contrary, are symmetric at all wall distances considered. Nevertheless, the non-Gaussian behaviour reflects also in the azimuthal velocity component, where the PDF in the vicinity of the wall ($y^+ = 2$) is flatter than further away from the wall[11]. Finally, the wall-normal velocity PDFs skewness varies while its flatness monotonically decreases when moving away from the wall (figure 4.8c). In general the mean and the variance of a PDF converge faster with the number of samples or the averaging interval than higher-order moments of the

[11] Note that a Gaussian distribution is symmetric about its mean and exhibits a constant flatness value of $F = 3$.

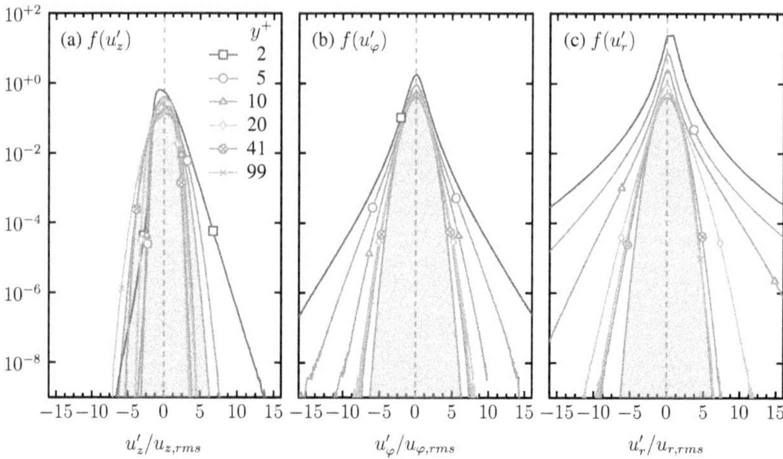

Figure 4.8: Probability density functions (PDF) of the fluctuating velocity components u'_i for case P180 at different wall-normal locations y^+. (a) $i=z$; (b) $i=\varphi$; (c) $i=r$. Grey shaded areas indicate a Gaussian distribution.

distribution, such as the skewness or the flatness. As discussed by Xu et al. (1996), large wall-normal velocities contribute to the high flatness level in the immediate vicinity of the wall. Since these strong events are rare both in space and time, a long sampling interval and/or a large domain size is required to obtain fully converged turbulence statistics. To evaluate the convergence of different statistical moments of the velocity distribution, these quantities are computed for P180 involving a very large sampling time ($\Delta T^+ \approx 350\,000$). A statistical quantity is considered as converged if it deviates less than one percent from the final value of P180. Subsequently, two additional simulations involving a finer (P180F) and a coarser grid (P180C) are carried out. For these DNSs statistics are integrated in time at least until the convergence criterion obtained from P180 is fulfilled. Statistically converged quantities obtained in the three DNSs are then compared to determine whether there is an influence of the near-wall grid refinement on turbulence statistics. Figure 4.9 shows the streamwise Reynolds stress at $y^+ = 15$ (figure 4.9a), the wall-normal skewness at $y^+ = 13.5$ (figure 4.9b) and the wall normal flatness in the vicinity of the wall (figure 4.9c) integrated over different time intervals for P180, P180C and P180F [12]. The peak of the streamwise

[12]Note that the wall-normal location, where the quantities are computed, correspond to prominent locations in the wall-normal profile of these quantities, see e.g. figure 4.10.

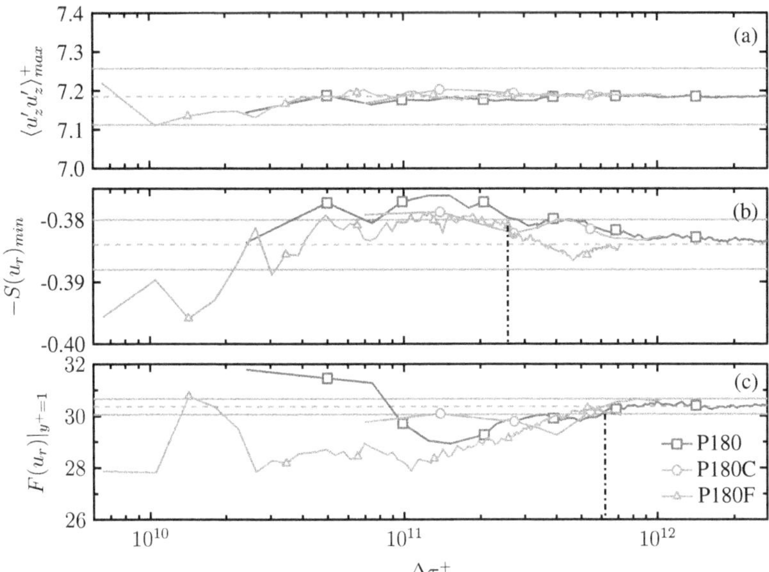

Figure 4.9: Convergence of Reynolds stresses and high-order statistical moments for different wall-normal grid spacings. (a) $\langle u'_z u'_z \rangle^+$ at $y^+ = 15$. (b) $S(u_r)$ at $y^+ = 13.5$. (c) $F(u_r)$ at at $y^+ = 1$. -□-, P180; -○-, P180C; -△-, P180F. Dashed gray lines, converged value for P180; solid gray lines, one percent criterion; dashed-dotted lines, integration time required for fulfilling the one percent criterion.

Reynolds stress component, depicted in figure 4.9(a), already exhibits a value within the one-percent criterion at the first evaluation time ($\Delta T^+ = 700$, $\Delta\tau^+ \approx 6\cdot 10^9$). In contrast, the skewness stays within the boundaries of the one-percent criterion after being integrated over $\Delta T^+ \approx 30000$ ($\Delta\tau^+ \approx 2.6\cdot 10^{11}$, figure 4.9b) and the high flatness level at $y^+ = 1$ requires $\Delta T^+ > 72500$ ($\Delta\tau^+ > 6.2\cdot 10^{11}$ to converge (figure 4.9c).

Another possibility of estimating the statistical convergence intervals for different order moments of a PDF is given by the following analytical considerations. According to Tennekes and Wyngaard (1972), the mean square relative error of a velocity component u_i due to finite sampling periods can be estimated by

$$\epsilon^2(\langle u_i\rangle_{\Delta T}) = \frac{\overline{(u_i-\overline{u_i})^2}}{\overline{u_i}^2}\frac{2\mathcal{T}}{\Delta T}, \tag{4.2}$$

when the integral time scale $\mathcal{T}$ is much smaller than the sampling period ΔT. The overbar denotes the true mean, which is obtained by averaging over an infinite time period ($\overline{u_i} = \int_0^\infty u_i\,\mathrm{d}t$), whereas the angular brackets with subscript ΔT denote averaging over the finite time period ΔT. The integral time scale is obtained by solving the integral $\mathcal{T} = \int_0^\infty R_{u_iu_i}(q)\,\mathrm{d}q$ with $R_{ii}(q) = \overline{u_i(t)u_i(t+q)}/\overline{u_i(t)^2}$, which is the temporal autocorrelation of u_i. We expand equation (4.2) in terms of two-dimensional spatial averaging and apply it to the variance ($\langle u_i'u_i'\rangle$) and the central fourth order moment $\langle u_i'^4\rangle$ of a velocity component u_i, yielding

$$\epsilon^2(\langle u_i'u_i'\rangle) = \frac{8\mathcal{T}_2^{(i)}\mathcal{L}_{z2}^{(i)}\mathcal{L}_{\varphi 2}^{(i)}}{\Delta T_2 L 2\pi r}\left(\frac{\overline{u_i'^4}}{\overline{u_i'^2}^2}-1\right), \tag{4.3}$$

$$\epsilon^2(\langle u_i'^4\rangle) = \frac{8\mathcal{T}_4^{(i)}\mathcal{L}_{z4}^{(i)}\mathcal{L}_{\varphi 4}^{(i)}}{\Delta T_4 L 2\pi r}\left(\frac{\overline{u_i'^8}}{\overline{u_i'^4}^2}-1\right), \tag{4.4}$$

where

$$\mathcal{T}_j^{(i)} = \int_0^\infty \frac{\overline{u_i'^j(t)u_i'^j(t+q)} - \overline{u_i'^j}^2}{\overline{u_i'^{2j}} - \overline{u_i'^j}^2}\,\mathrm{d}q, \tag{4.5}$$

$$\mathcal{L}_{kj}^{(i)} = \int_0^\infty \frac{\overline{u_i'^j(k)u'^j(k+\Delta k)} - \overline{u_i'^j}^2}{\overline{u_i'^{2j}} - \overline{u_i'^j}^2}\,\mathrm{d}\Delta k, \text{ with } k = z, \varphi, \tag{4.6}$$

are the integral time and length scales of $u_i'^j$, respectively. Note that the prime symbols in equations (4.3) and (4.4) denote the fluctuation about the true mean ($u_i' = u_i - \overline{u_i}$).

Substituting the inversion of the equations (4.3) and (4.4) into equation (4.1) results in an estimate of the spatio-temporal averaging interval needed to obtain a certain relative mean square error,

$$\Delta\tau_{i,2}^{+} = \frac{8\mathcal{T}_{2}^{(i)+}\mathcal{L}_{z2}^{(i)+}\mathcal{L}_{\varphi 2}^{(i)+}}{\epsilon^{2}}\left(\frac{\overline{u_i'^4}}{\overline{u_i'^2}^2}-1\right), \quad (4.7)$$

$$\Delta\tau_{i,4}^{+} = \frac{8\mathcal{T}_{4}^{(i)+}\mathcal{L}_{z4}^{(i)+}\mathcal{L}_{\varphi 4}^{(i)+}}{\epsilon^{2}}\left(\frac{\overline{u_i'^8}}{\overline{u_i'^4}^2}-1\right). \quad (4.8)$$

Inserting a relative error of one percent ($\epsilon = 0.01$), the above equations are evaluated at prominent wall-normal locations for the statistical quantities of interest, leading to the estimate of the spatio-temporal averaging interval. Note that the values on the right hand side of equations (4.7) and (4.8) are estimated by means of finite spatio-temporal averaging intervals themselves (i.e. $\overline{u_i} \approx \langle u_i \rangle$). Moreover, integral length and time scales, introduced in equations (4.5) and (4.6), are approximated by integrating the corresponding autocorrelation functions up to the first zero crossing (O'Neill et al., 2004). The computation yields values of $\Delta\tau_{z,2}^{+}(y^+ = 15) = 1.1 \cdot 10^9 \sim \mathcal{O}(10^9)$ and $\Delta\tau_{r,4}^{+}(y^+ = 1) = 1.2 \cdot 10^{11} \sim \mathcal{O}(10^{11})$. The large discrepancy between the averaging interval needed for the quantities above to reach the same level of statistical convergence is caused by two competing effects. On the one hand, the required averaging time or area increases with integral time or length scales. While the integral scales of $\langle u_z' u_z' \rangle$ at $y^+ = 15$ are equal to $\mathcal{T}_2^{(z)+} = 8$, $\mathcal{L}_{z,2}^{(z)+} = 110$ and $\mathcal{L}_{\varphi,2}^{(z)+} = 13$, the integral scales of $\langle u_r'^4 \rangle$ at $y^+ = 1$ are much smaller ($\mathcal{T}_4^{(r)+} = 1.2$, $\mathcal{L}_{z,4}^{(r)+} = 11.5$ and $\mathcal{L}_{\varphi,4}^{(r)+} = 6.8$). On the other hand, high-order moments appearing in the right hand side of equations (4.7) and (4.8) are larger ($F(u_z)|_{y^+=15} = 2.2$, $(\langle u_z'^8 \rangle / \langle u_z'^4 \rangle^2)|_{y^+=15} = 4.8$) than lower order terms and the strong intermittent behaviour of the wall-normal velocity at the wall leads to much larger values than the streamwise velocity component ($(\langle u_z'^8 \rangle / \langle u_z'^4 \rangle^2)|_{y^+=1} = 85$, $(\langle u_r'^8 \rangle / \langle u_r'^4 \rangle^2)|_{y^+=15} = 1.7 \cdot 10^4$). Cumulatively, the latter effect weighs more heavily, leading to much larger sampling intervals required for the convergence of the wall-normal flatness than for the streamwise variance.

The effective intervals required to obtain statistical convergence (see figure 4.9) are larger than the ones estimated by equations (4.7) and (4.8), albeit they are of the same order of magnitude ($\mathcal{O}(10^9)$ for the streamwise Reynolds stress and $\mathcal{O}(10^{11})$ for the wall-normal flatness). Consequently, the flow variables of all different Reynolds number simulation cases, presented in table 4.1, are integrated over at least $\Delta\tau^+ = 6.2 \cdot 10^{11}$.

Note that the analysis in section 4.2.2 particularly focuses on the interaction of outer-flow scales with the near-wall quantities, which becomes more important as the Reynolds number increases. As a consequence, the averaging interval needed for statistical convergence of a turbulence quantity in the vicinity of the wall is not supposed to scale in wall units solely, when considering high Reynolds numbers. Thus, the integration interval required to obtain convergence of a bulk variable is estimated from the bulk averaging interval evolution of the above quantities for P180 at $y/R = 0.8$. An averaging interval of $\Delta\tau_b \approx 16000$ ($\Delta T_b \approx 300$) was found to be sufficient (figure D.1 in the appendix) and serves together with the spatio-temporal interval estimated in wall units $\Delta\tau^+$, which is reported above, as a minimum limit for statistical convergence.

Next, the convergence of the wall-normal grid spacing in the vicinity of the wall is considered for the statistically converged quantities. Figure 4.10 shows a comparison of statistically fully converged profiles of the Reynolds stresses (figure 4.10a), the velocity skewness (figure 4.10b) and the velocity flatness (figure 4.10c) in the vicinity of the wall between P180, P180C and P180F. In addition to my data, experimentally measured profiles of the wall-normal skewness (figure 4.10b) and the wall-normal flatness (figure 4.10c) by Durst et al. (1995b) are added to the figure. While all profiles collapse well between P180, P180C, and P180F, indicating there is no influence of the grid spacing up to $\Delta r^+_{min} = 0.5$, the numerically obtained wall-normal flatness profiles exhibit a large discrepancy with respect to the experimentally measured profile. The lack of the large wall-normal flatness value in the vicinity of the wall in experimental data is due to highly intermittent wall-normal velocity fluctuations (velocity spikes), introduced in section 1.2.5 and further discussed hereinafter in section 4.3.1. Applying the Taylor expansion to the velocity components at the wall, the asymptotic limit of the statistical moments as they approach the wall ($r \to R$) can be derived. These asymptotes are plotted as grey lines in figure 4.10. The Reynolds stress profiles computed from the current data match the analytically derived expressions

$$\begin{aligned} &\lim_{r\to R} \langle u'_z u'_z \rangle \sim y^2, \\ &\lim_{r\to R} \langle u'_\varphi u'_\varphi \rangle \sim y^2, \\ &\lim_{r\to R} \langle u'_r u'_r \rangle \sim y^4, \end{aligned} \tag{4.9}$$

well. All components of the velocity skewness and flatness asymptotically approach a constant value, as they reach the wall. Figure 4.10(d) indicates that for P180 and P180C a constant value of wall-normal skewness has not been reached yet at the wall. Moreover, regarding the wall-normal flatness (4.10), where the gradients close to the

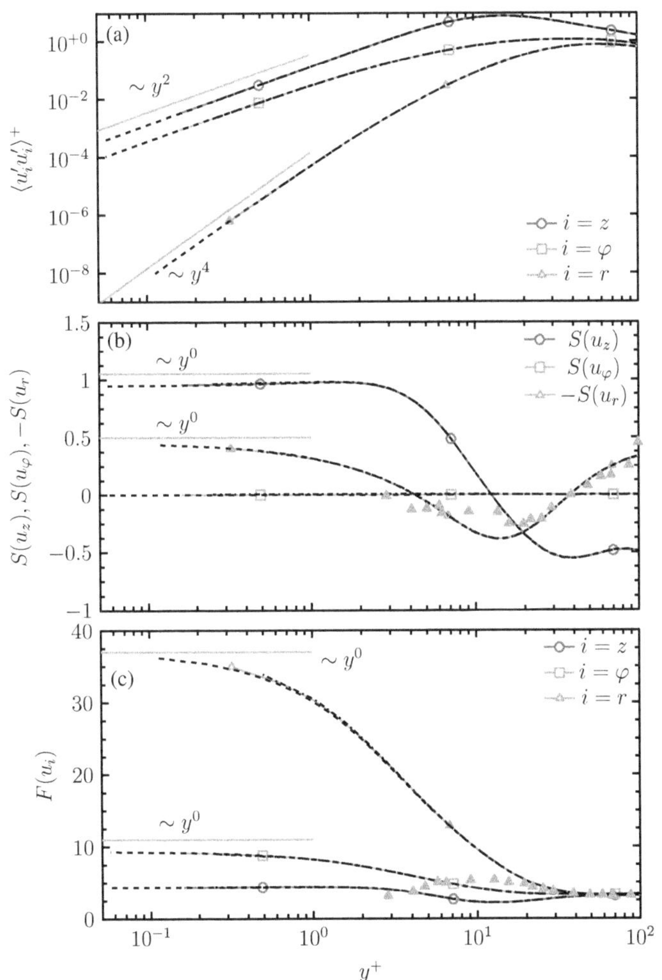

Figure 4.10: Near-wall statistical moments of the PDF of P180 (solid lines), P180F (dashed black lines) and P180C (dashed-dotted black lines). Note that the curves of the different cases collapse. ▲, LDA data of Durst et al. (1995a). (a) Reynolds stresses $\langle u_i' u_i' \rangle^+$. (b) Skewness $S(u_i)$. (c) Flatness $F(u_i)$.

wall are much steeper, none of the three simulations reaches a constant value. However, the separation between the three curves is very small, which leads to the conclusion that in the vicinity of the wall a grid sizing of $\Delta r^+_{min} \approx 0.5$ is sufficiently small for the wall-normal flatness to fulfil the boundary condition correctly.

4.2.2 Scaling of One-point Statistics

In the following, the Reynolds number dependency of one-point turbulence statistics is investigated by means of turbulent pipe flow DNSs of Reynolds numbers $180 \leq \mathrm{Re}_\tau \leq 1500$. As postulated in section 4.2.1, statistics are obtained by integrating flow variables over at least $\Delta\tau^+ = 2.6 \times 10^{11}$ and $\Delta T_b = 300$. Note that the latter criterion is not fulfilled by the simulation case P2880 and particularly high-order statistical moments might not be converged for this case. Thus, results obtained from P2880 are not included in the scaling analysis, although they are reported here for comparison.

Relation between Re_τ and Re_b

By approximating the mean velocity profile over the full wall-normal distance with the log law — equation (1.52) —, the following relation between the friction Reynolds number $\mathrm{Re}_\tau = u_\tau h/\nu$ and the bulk Reynolds number $\mathrm{Re}_b = u_b h/\nu$ can be derived for wall-bounded turbulence,

$$\frac{\mathrm{Re}_b}{\mathrm{Re}_\tau} = \frac{1}{\kappa}(\ln(\mathrm{Re}_\tau) - 1) + B, \tag{4.10}$$

whose approximation

$$\mathrm{Re}_\tau = 0.166\mathrm{Re}_b^{0.88} \tag{4.11}$$

has been shown to be valid for plane channel flows, with a channel half height h, over a range of Reynolds numbers (Hoyas and Jiménez, 2006; Lee and Moser, 2015; Moser et al., 1999). For pipe flow, the bulk Reynolds number is typically computed with the pipe diameter, i.e. $\mathrm{Re}_b = u_b D/\nu$. Since the equivalent length to the channel half height h or the boundary layer thickness δ in turbulent pipe flow is the pipe radius R, the approximation of equation (4.10) for turbulent pipe flow is computed with the bulk Reynolds number based on the radius ($\mathrm{Re}_{b,R} = \mathrm{Re}_b/2$). Fitting the approximation with the DNS data of the current study and other pipe flow DNS studies (Ahn et al.,

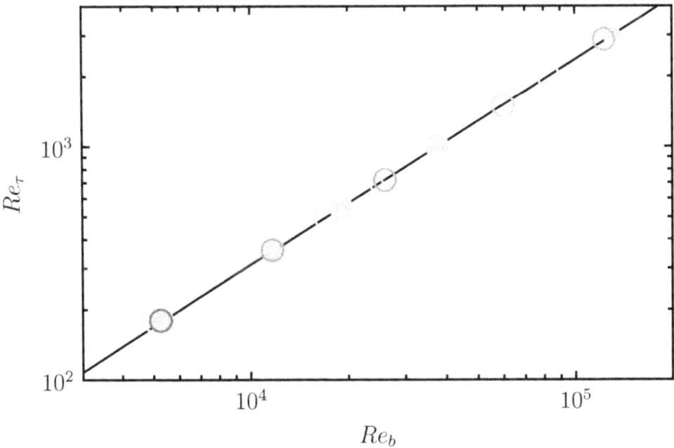

Figure 4.11: Variation of the friction Reynolds number Re_τ as a function of the bulk Reynolds number Re_b. P180 (○), P360 (○), P720 (○), P1500 (○), P2880 (○). DNS data from El Khoury et al. (2013) ( ), Wu et al. (2012) ( ), Chin et al. (2014) ( ), Ahn et al. (2015) ( ). The solid line corresponds to the power law $\mathrm{Re}_\tau = 0.173(\mathrm{Re}_b/2)^{0.88}$.

2015; Chin et al., 2014; El Khoury et al., 2013; Wu et al., 2012) yields

$$\mathrm{Re}_\tau = 0.173\mathrm{Re}_{b,R}^{0.88}. \tag{4.12}$$

Note that relation (4.12) for turbulent pipe flow differs slightly from the relation of turbulent plane channel flow (4.11), which is presumably caused by the difference in the mean velocity profile in the outer layer between these types of flow. Figure 4.11 shows the variation of the friction Reynolds number Re_τ as a function of the bulk Reynolds number Re_b for different turbulent pipe flow simulations of the current study and literature data together with equation (4.12). The above power law can be used to estimate the mass flow rate required to obtain a certain Re_τ, when setting up experiments or numerical simulations.

Mean Velocity Profile

The mean streamwise velocity profiles for different Re_τ are depicted in figure 4.12 together with data from El Khoury et al. (2013). Where the Reynolds number matches between El Khoury et al. (2013) and my data — i.e. $\mathrm{Re}_\tau \in \{180, 360\}$, the profiles collapse well, showing the validity of the current data. Figure 4.13 features a logarithmic

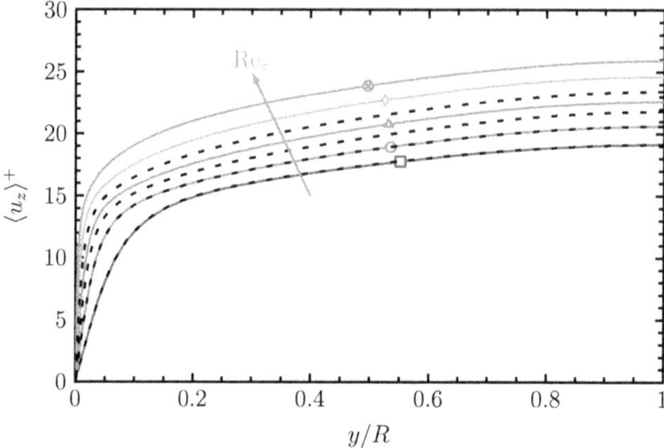

Figure 4.12: Mean streamwise velocity profile $\langle u_z \rangle^+$ as a function of the wall distance y/R. (a), y-coordinate linear in bulk units; (b), y-coordinate logarithmic in wall units. -□-, P180; -○-, P360; -△-, P720; -◇-, P1500; -⊗-, P2880; - - -, DNS data from El Khoury et al. (2013) with $\mathrm{Re}_\tau \in \{180, 360, 550, 1000\}$.

representation of the mean velocity profiles, which are found to follow the *linear law of the wall* — equation (1.51) — as well as the *logarithmic law of the wall* — equation 1.52. Regarding the ongoing debate whether a log law or a power law is valid for the high-Reynolds-number limit (see section 1.1.2), the Reynolds numbers considered in the current work are too low to ultimately address this question[13]. However, the validity of both log law and power law for the current regime of Reynolds numbers is verified in the following. Taking the wall-normal derivative of the log law (1.52) and multiplying the result with the wall-normal coordinate y^+ results in

$$y^+ \frac{\mathrm{d}\langle u_z \rangle^+}{\mathrm{d}y^+} = \frac{1}{\kappa}, \tag{4.13}$$

whereas taking the wall-normal derivative of the power law (1.2) and multiplying the result with the $y^+/\langle u_z \rangle^+$ results in

$$\frac{y^+}{\langle u_z \rangle^+} \frac{\mathrm{d}\langle u_z \rangle^+}{\mathrm{d}y^+} = \alpha. \tag{4.14}$$

[13]Moreover, friction Reynolds number required to answer the question will not be achievable by DNSs in the foreseeable future, since high-Reynolds-number pipe flow experiments with $\mathrm{Re}_\tau = \mathcal{O}(10^5)$ were not able to ultimately answer the question so far.

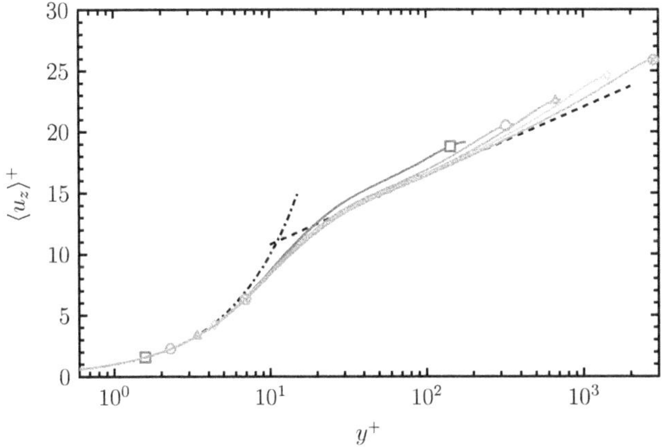

Figure 4.13: Mean streamwise velocity profile $\langle u_z \rangle^+$ as a function of the wall distance in wall units y^+ and logarithmic representation. -□-, P180; -○-, P360; -△-, P720; -◇-, P1500; -⊕-, P2880; - - -, linear law of the wall $\langle u_z \rangle^+ = y^+$; - - -, von Kármán log law: $\langle u_z \rangle^+ = 1/0.41 \log(y^+) + 5.2$.

Since the region where the function on the left-hand side of expression (4.13) — or the left-hand side of expression (4.14) — is approximately constant indicates the validity of the log law — or the power law —, these functions are called indicator functions. Log law and power law indicator functions are depicted in figure 4.14. While the profiles of the log law indicator function do not exhibit a plateau for the Reynolds number considered (figure 4.14), the power law indicator function profile of the largest Reynolds number considered ($\mathrm{Re}_\tau = 2880$) shows a plateau at $70 \lesssim y^+ \lesssim 1000$. Consequently, the present data suggests the validity of the power law at $\mathrm{Re}_\tau = 2880$. Both the local minimum in the log law indicator function with a value of $1/0.43$ and the plateau in the power law indicator function with a value of 0.145 match values of turbulent pipe flow DNS at $\mathrm{Re}_\tau = 3008$ reported by Ahn et al. (2015) as well as experimental data at $5383 \leq \mathrm{Re}_\tau \leq 39935$ reported by Willert et al. (2017). Moreover, the latter study also reports on a deviation from the plateau in the power law indicator function for $y^+ > 200$ and an approximate plateau in the log law indicator function for $200 < y^+ < 2000$, and thus, suggests the validity of the log law for Reynolds numbers larger than the ones considered here.

Moreover, taking into consideration the effect of the pressure gradient on the mean velocity profile by means of a higher-oder perturbation, as done by Luchini (2017),

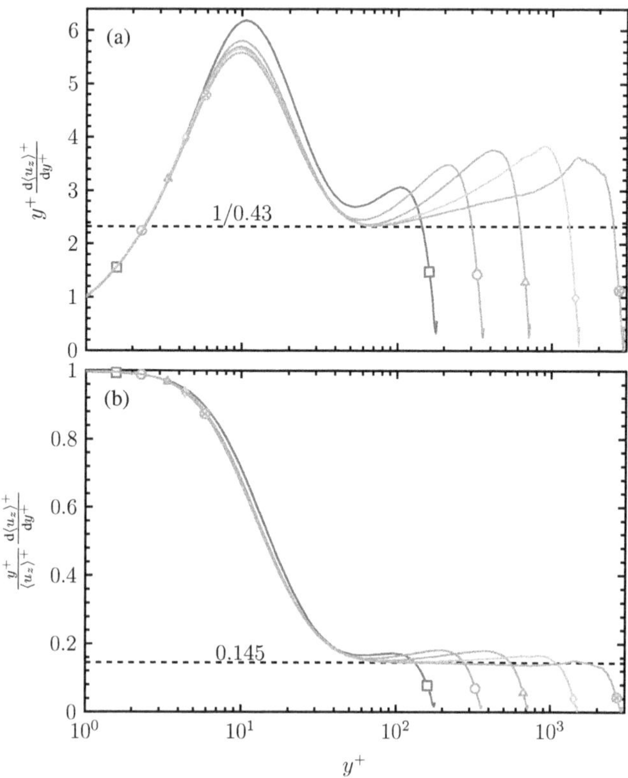

Figure 4.14: Indicator functions for the log law $y^+(\mathrm{d}\langle u_z\rangle^+/\mathrm{d}y^+)$ (a) and the power law $(y^+/\langle u_z\rangle^+)\cdot(\mathrm{d}\langle u_z\rangle^+/\mathrm{d}y^+)$ (b). -□-, P180; -○-, P360; -△-, P720; -◇-, P1500; -⊗-, P2880. Dashed lines indicate the local minimum (a) or the plateau region (b), respectively.

results in an alternative formulation of the log law,

$$\langle u_z \rangle^+ = \frac{1}{\kappa} \log y^+ + \frac{A_1 g}{\mathrm{Re}_\tau} y^+ + B, \tag{4.15}$$

with A_1 being a new universal constant that Luchini (2017) estimated as $A_1 \approx 1$ and $g = -R/\tau_w(\partial p/\partial z)$ being a geometry parameter ($g = 0$ for turbulent Couette flow, $g = 1$ for turbulent plane channel flow, and $g = 2$ for turbulent pipe flow). Subtracting the mean pressure gradient influence from the mean velocity leads to the profiles presented in figure 4.15. In contrast to the uncorrected mean velocity profiles displayed in figure 4.13, the corrected mean velocity profiles in figure 4.15(a) follow the log law almost up to the pipe axis. Consequently, the corresponding indicator function presented in figure 4.15(b) is much closer to a plateau than the one for the uncorrected data (figure 4.14a). In the high-Reynolds-number limit, the second term on the right-hand-side of expression (4.15) becomes negligibly small and the modified log law becomes the log law as developed by von Kármán. Besides, Luchini (2017) showed that the influence of the geometry on the mean velocity profile disappears when taking into account the effect of the pressure gradient. Thus, the modified log law in equation 4.15 shows universal behaviour.

Reynolds Stresses

The Reynolds stresses are second-order moments of the velocity distribution and the components of the Reynolds stress tensor **R**, which for turbulent pipe flow with homogeneous directions z and φ reads

$$\mathbf{R} = \begin{bmatrix} \langle u_z' u_z' \rangle & 0 & \langle u_z' u_r' \rangle \\ 0 & \langle u_\varphi' u_\varphi' \rangle & 0 \\ \langle u_r' u_z' \rangle & 0 & \langle u_r' u_r' \rangle \end{bmatrix}, \tag{4.16}$$

where $\langle u_z' u_r' \rangle = \langle u_z' u_r' \rangle$ due to symmetry. The diagonal components are normal Reynolds stresses and contribute to the turbulent kinetic energy

$$k = \frac{1}{2} \langle u_i' u_i' \rangle = \frac{1}{2} \left(\langle u_z' u_z' \rangle + \langle u_\varphi' u_\varphi' \rangle + \langle u_r' u_r' \rangle \right), \tag{4.17}$$

whereas the off-diagonal component $\langle u_z' u_r' \rangle$ is a shear stress. The square roots of the normal Reynolds stresses are the RMS values of the velocity fluctuations

$$u_{i,rms} = \sqrt{\langle u_i' u_i' \rangle} \text{ for } i = z, \varphi, r, \tag{4.18}$$

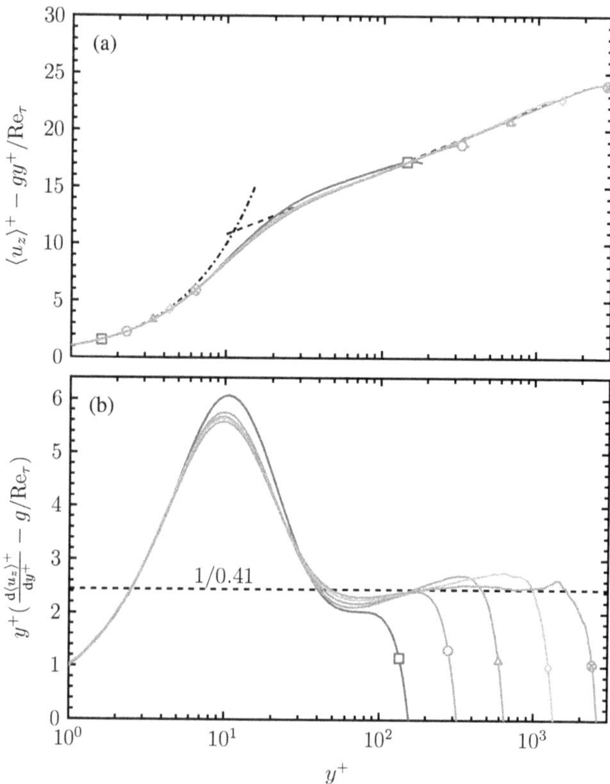

Figure 4.15: Modified mean velocity profile according to Luchini (2017) (a) and the corresponding indicator function $y^+(\mathrm{d}\langle u_z \rangle^+/\mathrm{d}y^+ - g/\mathrm{Re}_\tau)$ (b). -□-, P180; -○-, P360; -△-, P720; -◇-, P1500; -⊗-, P2880. (a) Dashed and dashed-dotted lines as in figure 4.13.

which are also referred to as turbulence intensities. Figure 4.16 shows the normal Reynolds stresses for the different Reynolds numbers considered over the wall-normal coordinate normalised in bulk units (left column) and in wall units (right column). When plotted against y/R, the streamwise Reynolds stress profiles collapse for $y/R \gtrsim 0.5$ and $\mathrm{Re}_\tau \leq 720$ (figure 4.16a). For higher Reynolds numbers, the profiles in the outer-flow region shift towards higher values collapsing for $y/R \gtrsim 0.5$ and $\mathrm{Re}_\tau \geq 1500$. The other two normal Reynolds stress components, displayed in figure 4.16(c,d), on the other hand, asymptotically approach Reynolds number independent profiles in the outer-flow region as the Reynolds number increases. Finally, the profiles of the azimuthal Reynolds stress component widely collapse for $y/R \gtrsim 0.1$ and $\mathrm{Re}_\tau \leq 720$ (figure 4.16c) and the profiles of the radial Reynolds stress component widely collapse for $y/R \gtrsim 0.1$ and $\mathrm{Re}_\tau \leq 1500$ (figure 4.16c). The normal Reynolds stress profiles plotted against y^+ are shown in figure 4.16(b,d,f). For the streamwise component, the profiles depend on the Reynolds number not only in the outer-flow region, but also in the buffer layer (figure 4.16b). Particularly, the streamwise Reynolds stress peaks around $y^+ = 15$ do not collapse for the different Reynolds numbers. As depicted in figure 4.16(d), the azimuthal Reynolds stress profiles do not collapse when plotted against y^+ similar to the streamwise Reynolds stress profiles. The radial Reynolds stress profiles presented in figure 4.16(f), on the contrary, collapse for $y^+ \lesssim 50$ and $\mathrm{Re}_\tau \geq 720$, while asymptotically approaching a plateau in the outer-flow region.

In the following, the Reynolds stress profiles in the log layer are compared with the predictions of Townsend's attached eddy hypothesis for high-Reynolds-number wall-bounded turbulence — see equations (1.67) to (1.70). For the streamwise Reynolds stress component, a logarithmic variation — as predicted in equation (1.70) — is not visible in the range of Reynolds numbers considered here. For much higher Reynolds number, however, experimental data from Marusic et al. (2013) suggest a slope of $B_z \approx 1.26$ for equation (1.67). A tangential fit of the latter equation to the current data at $\mathrm{Re}_\tau = 2880$ with the $B_z \approx 1.26$ is, thus, shown in figure 4.16(b), for comparison. For the azimuthal Reynolds stress component, on the other hand, a logarithmic region is clearly visible at lower Reynolds numbers. Different fits of the spanwise Reynolds stress profile in this region to Townsend's prediction have been reported by Bernardini et al. (2014); Jiménez and Hoyas (2008); Lee and Moser (2015) for turbulent plane channel flow. For turbulent pipe flow of $\mathrm{Re}_\tau = 2003$, Chin et al. (2014) reported $\langle u'_\varphi u'_\varphi \rangle \approx 4.3 - 0.456 \log(y/R)$. A somewhat smaller slope is obtained from the current data, where $\langle u'_\varphi u'_\varphi \rangle \approx 0.86 - 0.43 \log(y/R)$ is the fit from the azimuthal Reynolds stress profile between $100 \leq y^+ \leq 400$ at $\mathrm{Re}_\tau = 2880$ on equation (1.68). As the grey

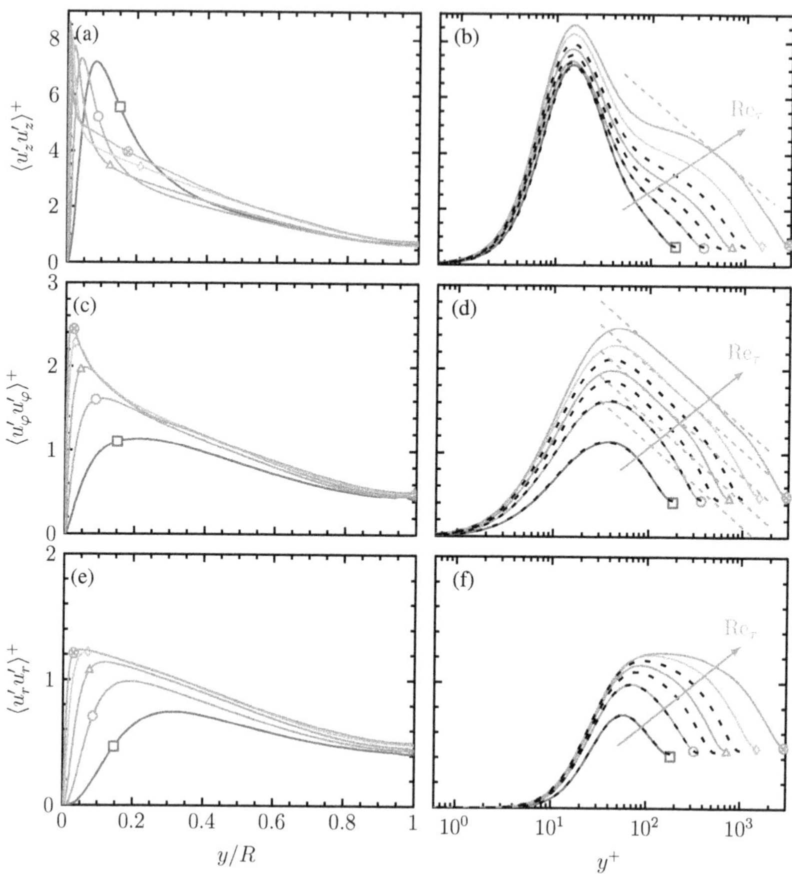

Figure 4.16: Normal Reynolds stress profiles in wall units $\langle u'_i u'_i \rangle^+$ for $i = z$ (a,b), $i = \varphi$ (c,d) and $i = r$ (e,f). (a,c,e), y-coordinate in bulk units; (b,d,f) y-coordinate in wall units. -□-, P180; -○-, P360; -△-, P720; -◇-, P1500; -⊗-, P2880; - - -, DNS data from El Khoury et al. (2013) with $\mathrm{Re}_\tau \in \{180, 360, 550, 1000\}$. Gray dashed lines indicate Townsend's predictions (cf. equations 1.67–1.70): (b) $1.78 - 1.26 \log(y/R)$; (c) $0.86 - 0.43 \log(y/R)$.

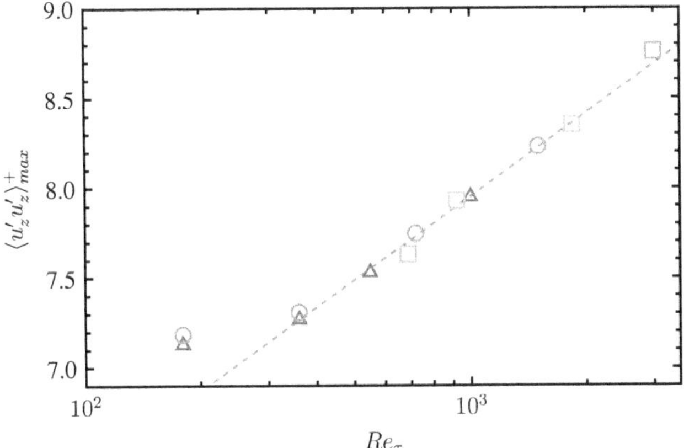

Figure 4.17: Variation of the streamwise Reynolds stress peak $\langle u'_z u'_z \rangle^+_{max}$ at $y^+ \approx 15$ with the Reynolds number Re_τ. ○, current DNS; △, El Khoury et al. (2013), □, Boersma (2013), □, Ahn et al. (2015).

dashed lines in figure 4.16(d) indicate, the logarithmic behaviour of the azimuthal Reynolds stress component is obtained for Reynolds numbers as low as $\mathrm{Re}_\tau = 720$. Similar to the streamwise Reynolds stress profile, the wall-normal profile — presented in figure 4.16(f) in inner scaling — does not exhibit universal scaling behaviour for the Reynolds numbers considered here. However, in agreement with equation (1.69), the profile tends to form a plateau when increasing the Reynolds number. Here, higher Reynolds number simulations are required to estimate the high-Reynolds-number limit constant A_r in equation (1.69).

The failure of wall scaling for the streamwise Reynolds stress peak at $y^+ = 15$ (figure 4.16b) is a known phenomenon and has been discussed for plane channel flow by Hoyas and Jiménez (2006) amongst others. Figure 4.17 shows the variation of the peak value with the Reynolds number for the current data and data from the literature. Lee and Moser (2015) fitted the logarithmic Reynolds number dependency of the streamwise Reynolds stress peak in plane channel flow with the expression $\langle u'u' \rangle^+_{max} = 0.642 \log(\mathrm{Re}_\tau) + 3.66$ for $180 \leq \mathrm{Re}_\tau \leq 5200$. However, unlike its channel counterpart, the peak value obtained from turbulent pipe flow does not follow a logarithmic scaling law over the full range of Reynolds numbers. For $\mathrm{Re}_\tau > 360$ the

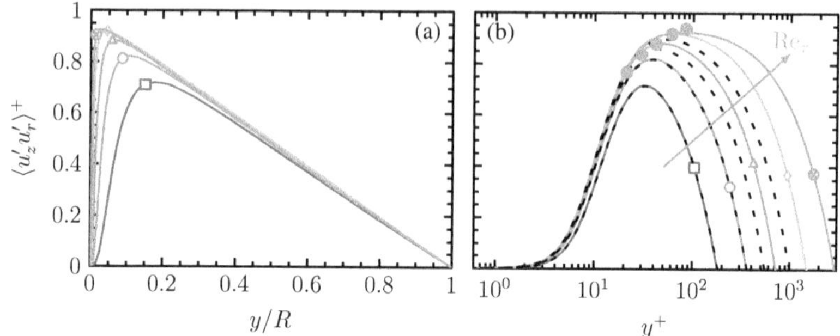

Figure 4.18: Reynolds shear stress profiles in wall units $\langle u_z' u_e' \rangle^+$. (a), y-coordinate in bulk units; (b) y-coordinate in wall units. -□-, P180; -○-, P360; -△-, P720; -◇-, P1500; -⊗-, P2880; - - -, DNS data from El Khoury et al. (2013) with $\mathrm{Re}_\tau \in \{180, 360, 550, 1000\}$; grey filled circles, estimated maxima.

logarithmic expression

$$\langle u_z' u_z' \rangle^+_{max} = 0.67 \log(\mathrm{Re}_\tau) + 3.324, \quad \mathrm{Re}_\tau > 360 \tag{4.19}$$

fits the dependency of the $\langle u_z' u_z' \rangle^+$-peak on Re_τ in turbulent pipe flow (dashed line in figure 4.17). The Reynolds number dependency of the streamwise Reynolds stress component and its relation to underlying coherent structures is further elaborated in section 4.4.

The only non-zero off-diagonal component of the Reynolds stress tensor in turbulent wall-bounded flows with two homogeneous directions is the Reynolds shear stress $\langle u_z' u_r' \rangle$. Figure 4.18 shows the variation of the Reynolds shear stress with the wall distance in bulk units and wall units, respectively. As indicated by figure 4.18(a), the maximum of the Reynolds shear stress increases and shifts towards the wall with increasing Reynolds number. While the contribution to the total shear stress is negligible in the very vicinity of the wall — where viscous forces play a major role —, it dominates the total shear stress in the outer flow. Following Townsend's prediction, the Reynolds shear stress profile tends to form a plateau of $\langle u_z' u_r' \rangle = 1$ in the logarithmic region for increasing Reynolds numbers, which is not yet obtained within the Reynolds numbers considered (figure 4.18b). Moreover, Lee and Moser (2015) deduced the Reynolds shear stress variation with the Reynolds number from the analytical expression for the total

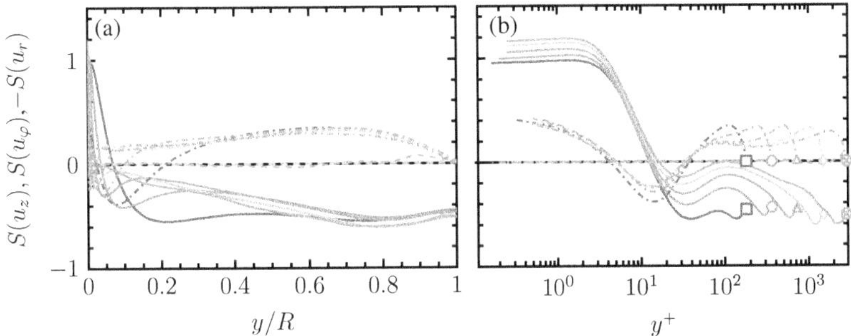

Figure 4.19: Velocity skewness profiles $S(u_i)$ with $i = z$ (solid lines), $i = \varphi$ (dashed lines), and $i = r$ (dashed-dotted lines). (a), y-coordinate in bulk units; (b) y-coordinate in wall units. -□-, P180; -○-, P360; -△-, P720; -◇-, P1500; -⊗-, P2880.

shear stress

$$\tau_{tot}^+ = \frac{\mathrm{d}\langle u_z\rangle^+}{\mathrm{d}y^+} + \langle u_z' u_r'\rangle^+ = 1 - y/R, \tag{4.20}$$

and the fact that the log law indicator function varies only little over a broad range $(y^+(\mathrm{d}\langle u_z\rangle^+/\mathrm{d}y^+) \approx 1/\kappa$ for $30 < y^+ < 0.75\mathrm{Re}_\tau$, cf. figure 4.14a). Thus, the Reynolds shear stress variation yields

$$\langle u_z' u_r'\rangle^+ \approx 1 - \frac{y^+}{\mathrm{Re}_\tau} - \frac{1}{\kappa y^+}, \quad \text{for } 30 < y^+ < 0.75\mathrm{Re}_\tau. \tag{4.21}$$

From expression (4.21) the maximum of the Reynolds shear stress can be estimated as $1 - 2/\sqrt{\kappa \mathrm{Re}_\tau}$ with its wall-normal location $y^+ \approx \sqrt{\mathrm{Re}_\tau/\kappa}$, which has been pointed out by several authors (Afzal, 1982; Lee and Moser, 2015; Morrison et al., 2004; Panton, 2007; Sillero et al., 2013). The predicted values for the current Reynolds numbers are indicated by grey filled symbols in figure 4.18(b). While the predictions for the lower Reynolds numbers, where a logarithmic region barely exists are poor, the maximum of the Reynolds shear stress is reasonably well estimated for $\mathrm{Re}_\tau \geq 720$.

Velocity Skewness

Figure 4.19 shows the velocity skewness profiles for the different Reynolds numbers. While the azimuthal profile equals to zero at all distances from the wall, the streamwise and wall-normal profiles reflect a non-Gaussian behaviour. Being positive at the solid boundary, the streamwise skewness $S(u_z)$ decreases when shifting away from the wall, crossing zero at $y^+ \approx 13$ and finally stays negative up to the pipe axis. The wall-normal skewness[14] $-S(u_r)$ is positive at the wall, then decreases until its minimum at $y^+ \approx 13$ (crossing zero at $y^+ \approx 5$), before crossing zero again at $y^+ \approx 34$ and reaching a local maximum at $y/R \approx 0.7$, whose location scales in outer length scale (figure 4.19a). While the wall-normal velocity skewness scales in wall units near the wall — apart from a low Reynolds number effect for $\mathrm{Re}_\tau = 180$ around $y^+ \approx 13$, figure 4.19(b) — and in bulk units in the outer-flow region (4.19a), the streamwise skewness profiles near the wall depend clearly on the Reynolds number. Comparing the streamwise skewness at two prominent locations among the different Reynolds numbers shows that the streamwise velocity distribution is larger positively skewed very close to the wall ($y^+ = 1.4$) and less negatively skewed at the location of the local streamwise skewness minimum at $y^+ \approx 33$, when increasing the Reynolds number. The Reynolds number dependency at both locations is fitted by the logarithmic expressions,

$$S(u_z)|_{y^+\approx 1.4} = 0.082 \log(\mathrm{Re}_\tau) + 0.548, \tag{4.22}$$

$$S(u_z)|_{y^+\approx 33} = 0.156 \log(\mathrm{Re}_\tau) - 1.343, \tag{4.23}$$

indicated by dashed lines in figure 4.20, where pipe flow DNS data of El Khoury et al. (2013) and experimental TBL data of Mathis et al. (2011) at $y^+ \approx 33$ is also shown for comparison. The above logarithmic scaling laws match well with the DNS data of El Khoury et al. (2013) and expression (4.23) predicts the values of the TBL data fairly well up to $\mathrm{Re}_\tau = 7300$.

Velocity Flatness

As shown in figure 4.21(a), all components of the velocity flatness are close to the Gaussian value of $F = 3$ throughout most of the flow domain ($y/R \gtrsim 0.2$), and thus, scale well with the Reynolds number. In the vicinity of the wall, on the contrary, a Reynolds number dependency appears primarily in the wall-normal flatness component (figure 4.21b). The already high level of wall-normal flatness at the wall, which is caused by velocity spikes (see section 4.1.1), shifts towards even higher values when increasing Re_τ. The variation of the wall-normal flatness at $y^+ = 1$ with the Reynolds number is displayed in figure 4.22, where also values from plane channel flow DNSs of

[14] For better comparability with other wall-bounded flows, where the wall-normal coordinate is usually pointing away from the wall, odd statistical moments of the wall-normal velocity component are given pointing in y-direction instead of r ($S(v) = -S(u_r)$).

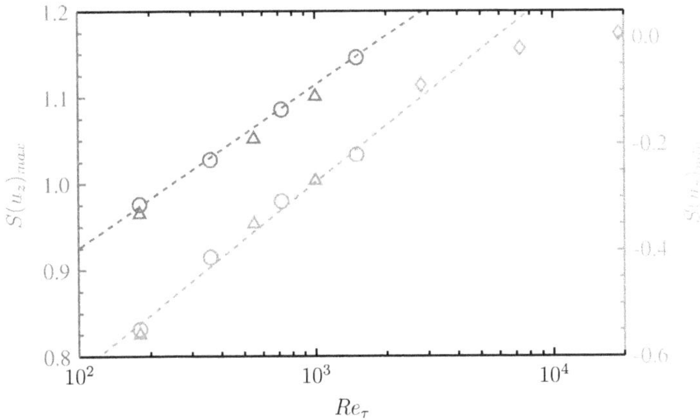

Figure 4.20: Variation of streamwise velocity skewness extrema with the Reynolds number Re_τ: $S(u_z)_{max}$ at $y^+ = 1.4$, $S(u_z)_{min}$ at $y^+ \approx 33$. Pipe flow DNS data: o, current DNS; $\triangle$, El Khoury et al. (2013). Experimental boundary layer data: $\diamond$, Mathis et al. (2011).

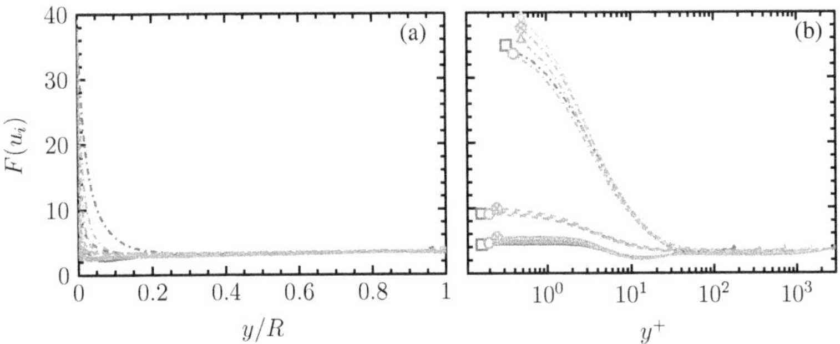

Figure 4.21: Velocity flatness profiles $F(u_i)$ with $i = z$ (solid lines), $i = \varphi$ (dashed lines), and $i = r$ (dashed-dotted lines). (a), y-coordinate in bulk units; (b) y-coordinate in wall units. -□-, P180; -○-, P360; -△-, P720; -◇-, P1500; -⊗-, P2880.

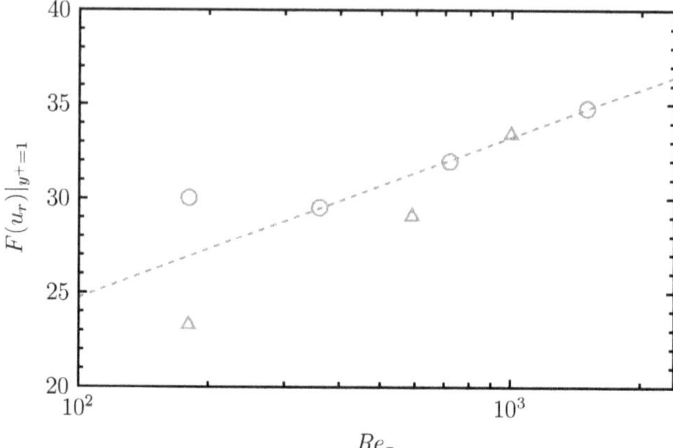

Figure 4.22: Variation of the wall-normal flatness $F(u_r)$ at $y^+ = 1$ with the Reynolds number Re_τ. ○, current pipe flow DNS data; △, plane channel flow DNS data by Lenaers et al. (2012).

Lenaers et al. (2012) are included[15]. For $\mathrm{Re}_\tau \geq 360$ the pipe flow data can be fitted with the logarithmic expression

$$3.7 \log(\mathrm{Re}_\tau) + 7.7, \quad \mathrm{Re}_\tau \geq 360, \tag{4.24}$$

indicated by the dashed line in figure 4.22. As for the streamwise Reynolds stress (figure 4.17), the wall-normal flatness in the vicinity of the wall (figure 4.22) of the lowest Reynolds number considered ($\mathrm{Re}_\tau = 180$) does not match with the corresponding logarithmic law for turbulent pipe flow. Moreover is the discrepancy between the values obtained from pipe flow and plane channel flow largest at this Reynolds number ($F(u_r)|_{y^+=1} = 30$ for pipe, $F(w)|_{y^+=1} = 23.3$ for channel flow, figure 4.22).

Vorticity

The intensity of vorticity fluctuations $\omega_{i,rms}$ is a measure for the activity of both vortical motions and shear fluctuations in turbulent flows. Figure 4.23 shows the variation of the vorticity fluctuation intensity in wall units $\omega^+_{i,rms}$ with y/R (figure 4.23a) and with

[15]Note that statistically converged near-wall high-order statistics of wall-bounded turbulence are rarely reported, which is due to either the computational effort (for numerical simulations, see section 4.2.1) or the accessibility of the near-wall region (for experimental measurements, especially in case of high Re_τ).

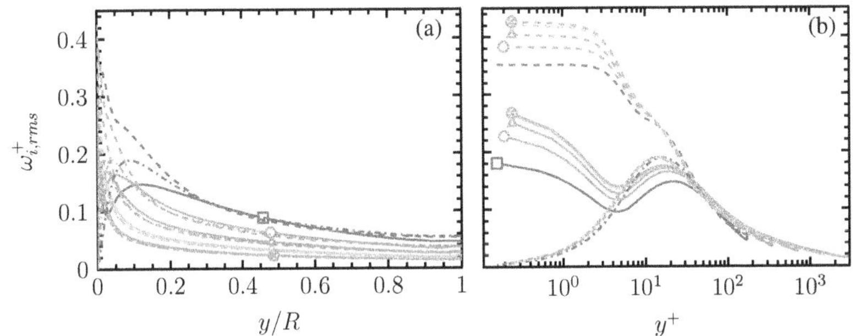

Figure 4.23: Vorticity fluctuation intensities in wall units $\omega^+_{i,rms}$ with $i = z$ (solid lines), $i = \varphi$ (dashed lines), and $i = r$ (dashed-dotted lines). (a), y-coordinate in bulk units; (b) y-coordinate in wall units. -□-, P180; -○-, P360; -△-, P720; -◇-, P1500; -●-, P2880.

y^+ (figure 4.23b). As the different Reynolds number vorticity fluctuation intensity component profiles presented in figure 4.23(a) do not collapse at any distance from the wall, the vorticity fluctuation intensity in wall units does not scale with the Reynolds number when plotted against y/R. When plotted over y^+, as shown in figure 4.23(b), however, the vorticity fluctuation intensity components appear to collapse within the log layer ($y^+ \gtrsim 30$, $y/R \lesssim 0.3$), independently of the Reynolds number. While being highly anisotropic in the vicinity of the no-slip boundary, the vorticity fluctuation intensity becomes isotropic above $y^+ \gtrsim 30$. In the vicinity of the wall, on the other hand, the wall-parallel vorticity intensities in wall units depend on the Reynolds number, whereas the wall-normal component scales well while being damped to zero by the no-slip condition. Unlike for velocity fluctuations, the region where wall-parallel vorticity fluctuation intensities in inner units are Reynolds number depended is restricted to a thin layer close to the wall ($y^+ \lesssim 10$, cf. figure 4.23). This observation is consistent with turbulent plane channel flow, where Hoyas and Jiménez (2008) found outer-flow VLSMs to be responsible for the Reynolds number dependency of wall-parallel vorticity fluctuation intensities. While they found energetic wave modes related to VLSM in two-dimensional pre-multiplied energy spectra at all wall distances they considered ($y^+ = 6, 10, 14$), they obtained these modes only from the enstrophy spectrum at $y^+ = 6$. Hoyas and Jiménez (2008) argued that these modes do generally not occur in enstrophy spectra, since they are irrotational. Very close to the wall, however, these modes need to fulfil the no-slip boundary condition, which results in a "long, wide and thin viscous layer" that reflects in the wall-parallel vorticity close to the wall ($y^+ < 10$).

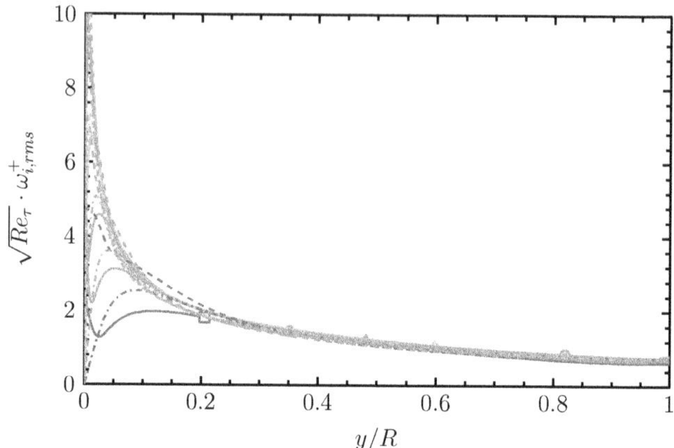

Figure 4.24: Vorticity fluctuation intensity $\omega^+_{i,rms}$ times $\sqrt{\mathrm{Re}_\tau}$ with $i = z$ (solid lines), $i = \varphi$ (dashed lines), and $i = r$ (dashed-dotted lines) plotted over y/R. -□-, P180; -○-, P360; -△-, P720; -◇-, P1500; -⊛-, P2880.

For the outer layer of wall-bounded turbulence, Antonia et al. (1996) and Panton (2009) showed that the vorticity fluctuation intensity acts mainly dissipative, and thus, scales with the Kolmogorov time scale $\omega_{i,rms} \sim \sqrt{(\nu R)/u_\tau}$. Consequently, vorticity fluctuation intensities multiplied by $\sqrt{\mathrm{Re}_\tau}$ after being normalised in wall units are plotted against y/R in figure 4.24. The collapse of the different Reynolds number pre-multiplied profiles presented in figure 4.24 indicates the scaling of the vorticity fluctuation intensity with $\sqrt{\mathrm{Re}_\tau}$ for the current data.

TKE Budget

This section describes the scaling behaviour of the TKE budget terms, which are introduced in equation (1.64). Figure 4.25 depicts the profiles of the TKE budget terms for the different Reynolds numbers, namely TKE production (solid lines), dissipation (dashed lines), turbulent diffusion (dashed-dotted lines), pressure diffusion (dotted lines), and viscous diffusion (loosely dashed lines). The variation of the TKE budget terms normalised in wall units with y/R in the outer layer is displayed in figure 4.25(a). As the profiles of the TKE production, dissipation and turbulent transport do not collapse for the different Reynolds numbers, scaling in wall units does not hold in the outer layer. Note that viscous diffusion and pressure diffusion are negligibly small

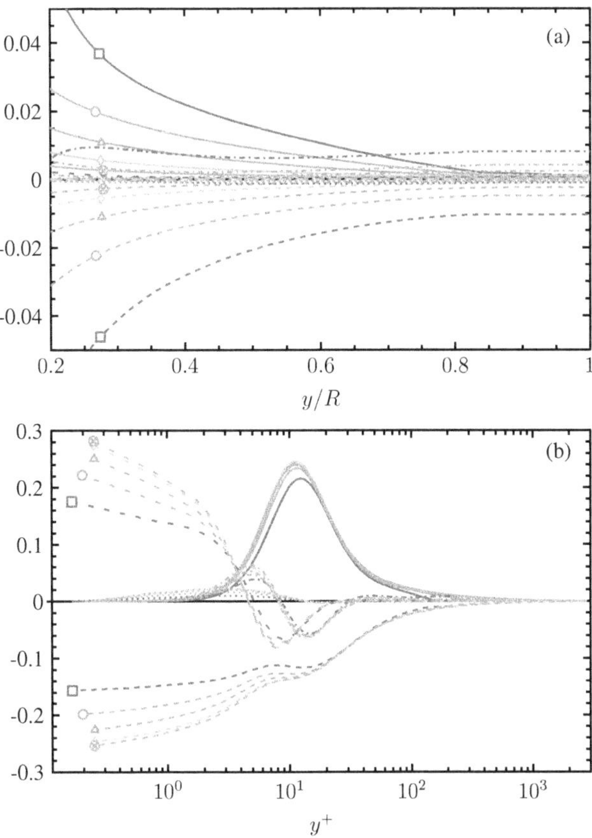

Figure 4.25: Profiles of the TKE budget terms normalised in wall units. (a), y-coordinate in bulk units; (b) y-coordinate in wall units. Solid lines, production ($\mathcal{P}$); dashed lines, pseudo-dissipation (ε); dashed-dotted lines, turbulent diffusion (TD); dotted, pressure diffusion (PD); loosely dashed, viscous diffusion (VD). -□-, P180; -○-, P360; -△-, P720; -◇-, P1500; -⊗-, P2880.

in the outer layer. In the inner layer above $y^+ \approx 10$, on the contrary, the different Reynolds number TKE budget term profiles collapse reasonably well when plotted against y^+ as depicted in figure 4.25(b). Very close to the wall ($y^+ \lesssim 10$), however, a Reynolds number dependency in the TKE budget terms is apparent. While the TKE production term only reflects a low Reynolds number effect and scales well for $\mathrm{Re}_\tau \geq 360$ in the vicinity of the wall (figure 4.25b, solid lines), the transport terms, as well as TKE dissipation, tend to larger values for higher Reynolds numbers over the full range of Reynolds numbers considered. As pointed out by Hoyas and Jiménez (2008) the scaling failure of TKE dissipation is related to the scaling failure of the wall-parallel vorticity fluctuation intensity discussed above. Since TKE dissipation is dominated by the enstrophy, its scaling failure is — as the scaling failure of the near-wall parallel vorticity fluctuation intensities — restricted to a thin layer close to the wall ($y^+ \lesssim 10$). Moreover, close to the wall, where viscous effects are dominant, the dissipation is mainly balanced by the viscous diffusion term. Consequently, the viscous diffusion term exhibits the same scaling failure as TKE dissipation (cf. figure 4.25b).

In the outer layer (which they defined above $y^+ \approx 150$), Hoyas and Jiménez (2008) found the TKE budget terms in turbulent plane channel flow to scale reasonably well with u_τ^3/h. The same scaling is reported for turbulent open channel flow by Calmet and Magnaudet (2003). Since the main contributors to TKE dissipation in the outer layer are small-scale vortices, and thus, the TKE dissipation scales with the enstrophy, Calmet and Magnaudet (2003) suggested to multiply the TKE budget terms with Re_τ after being normalised in wall units, which results in the same scaling as proposed by Hoyas and Jiménez (2008) for the plane channel flow case. Following the last mentioned authors, the TKE budget terms normalised by u_τ^3/R for turbulent pipe flow are presented in figure 4.26. As for the channel flow cases, the suggested scaling works reasonably well in the outer layer of turbulent pipe flow, where the profiles of TKE production, dissipation and turbulent transport collapse for different Re_τ. Note that the TKE production and transport profiles only collapse well for the larger Reynolds numbers ($\mathrm{Re}_\tau \geq 360$) and further away from the wall ($y/R \gtrsim 0.5$), whereas the TKE dissipation profile normalised by u_τ^3/R scales well for all Reynolds numbers throughout the outer layer ($y^+ \gtrsim 50$). This discrepancy might be caused by the fact that the scaling argument for the TKE dissipation is derived based on its link to the vorticity fluctuation intensity. As the vorticity fluctuation intensity scales well throughout the outer layer when normalised with the Kolmogorov time scale, so does the TKE dissipation. For the TKE production and turbulent transport, on the contrary, a contribution from scales much larger than the Kolmogorov scales is presumable. Thus,

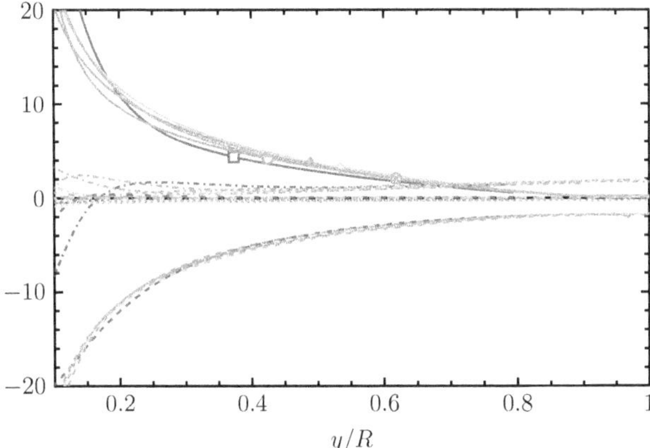

Figure 4.26: Profiles of the TKE budget terms normalised by u_τ^3/R. Solid lines, production ($\mathcal{P}$); dashed lines, pseudo-dissipation (ε); dashed-dotted lines, turbulent diffusion (TD); dotted, pressure diffusion (PD); loosely dashed, viscous diffusion (VD). -□-, P180; -○-, P360; -△-, P720; -◇-, P1500; -⊗-, P2880.

the scaling argument for the TKE dissipation does not fully hold for the other TKE budget terms in the outer layer.

4.3 Two-point Statistics

In the upcoming chapter, the scaling of coherent structures in turbulent pipe flow, particularly SSMs, LSMs, and VLSMs, is analysed by means of two-point statistical quantities — such as spectra and two-point correlations. Parts of the analysis have been published in Bauer et al. (2017). As introduced in section 1.2.5, coherent structures play an important role for the distribution of turbulent kinetic energy in wall-bounded turbulence. Well-known wall-layer streaks, which are elongated regions of either high or low streamwise momentum, contribute most to the formation of the $\langle u_z' u_z' \rangle$-peak. The mean spanwise spacing of velocity streaks can be extracted from two-point velocity correlations or pre-multiplied energy spectra. Kim et al. (1987) reported a value of $\lambda^+ \approx 100$ for plane channel flow, whereas Unger (1994) and Wagner and Friedrich (1998) reported $\lambda_\varphi^+ \approx 120$ for turbulent pipe flow. Moreover, Unger (1994) measured the streamwise extension of wall-layer streaks as approximately $\lambda_z^+ \approx 1000$. First, the scaling of velocity spikes is determined in section 4.3.1, before the scaling of wall-layer streaks and VLSMs is analysed by means of velocity correlations and pre-multiplied energy spectra in section 4.3.2.

4.3.1 Velocity Spikes

The dynamics of velocity spikes in turbulent pipe flow has been discussed in section 4.1 for a Reynolds number of $\mathrm{Re}_\tau = 180$. To analyse the scaling of velocity spikes, additional Reynolds numbers are taken into account and two-point velocity correlations of the wall-normal velocity in a wall-parallel plane

$$\rho_{rr}(\Delta z, \Delta\varphi, r_0) = \frac{\langle u_r'(z_0, \varphi_0, r_0) u_r'(z_0 + \Delta z, \varphi_0 + \Delta\varphi, r_0)\rangle}{u_{r,rms}'^2(r_0)}, \tag{4.25}$$

as well as modified two-point velocity correlations of the wall-normal velocity in a cross-sectional plane

$$\rho_{rr}^*(\Delta\varphi, \Delta r, r_0) = \frac{\langle u_r'(z_0, \varphi_0, r_0) u_r'(z_0, \varphi_0 + \Delta\varphi, r_0 + \Delta r)\rangle}{u_{r,rms}'^2(r_0 + \Delta r)}, \tag{4.26}$$

are computed. The modified correlation is normalised by the corresponding RMS value at the wall-normal separation length instead of the RMS value at the reference point. This normalisation has been chosen to point out the relevance of velocity spikes for the velocity flatness, since only large ratios of the velocity fluctuation and the local velocity RMS reflect in a large velocity flatness. Figure 4.27 shows iso-contours of two-dimensional two-point velocity correlations of the wall-normal velocity computed according to equation (4.25) and (4.26) two orthogonal planes with separation lengths in wall units. The iso-contours of the wall-normal velocity correlation at $y^+ = 2$ in the $z\varphi$-plane, displayed in figure 4.27(a), collapse well for the range of Reynolds number considered — with little deviations for the lower Reynolds numbers ($\mathrm{Re}_\tau \leq 360$). Additionally, the iso-contours of the modified wall-normal velocity correlation (4.26) in the pipe cross-section, presented in figure 4.27(b), collapse well for $\mathrm{Re}_\tau \geq 720$. The upper part of the smaller Reynolds numbers iso-contours, as well as the vertical location of the negative correlated regions (dashed lines), deviates somewhat from the profiles at $\mathrm{Re}_\tau \geq 720$. The latter is presumably caused by the pipe curvature, which increases in wall units when considering smaller Reynolds numbers. In summary, however, velocity spikes appear to scale well in wall units. Estimating their spatial extension via velocity correlation iso-contours results in

$$l_{z,spike}^+ \times l_{\varphi,spike}^+ \times l_{r,spike}^+ \approx 120 \times 25 \times 6, \tag{4.27}$$

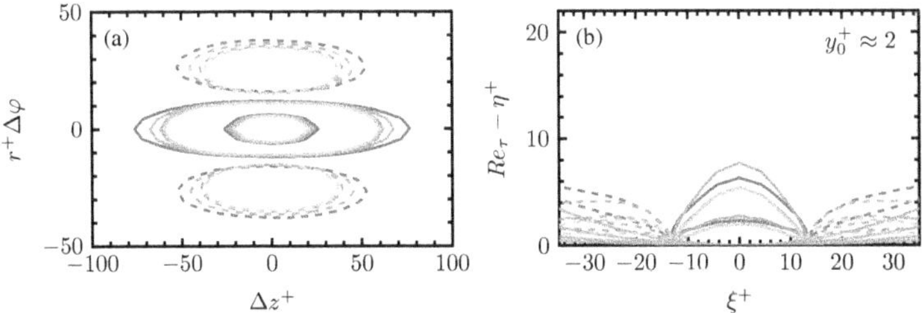

Figure 4.27: Two-dimensional velocity correlations of the streamwise velocity ρ_{rr} and ρ^*_{rr} with reference point at $y_0^+ = 2$. (a) $\rho_{rr}(\Delta z, \Delta\varphi, r)$; (b) $\rho^*_{rr}(\Delta\varphi, \Delta r, r)$. Cartesian cross-sectional coordinates $\xi = (r_0 + \Delta r)\sin(\Delta\varphi)$, $\eta = (r_0 + \Delta r)\cos(\Delta\varphi)$. Solid(dashed) iso-contours indicate values of +(−)0.12, 0.62. —, P180; —, P360; —, P720; —, P1500.

the latter being mostly independent of the Reynolds number.

4.3.2 Velocity Streaks and VLSMs

To determine the size of energetic coherent structures statistically, one usually looks at pre-multiplied energy spectra. The peak location in a pre-multiplied energy spectrum is considered as the wavelength of the most energy-containing motion. Figure 4.28 shows the streamwise one-dimensional pre-multiplied energy spectrum of the streamwise velocity component as a function of the streamwise wavelength and the wall distance both in wall units (figure 4.28a) and in bulk units (figure 4.28b)[16]. While the short wavelength part of the spectrum in the vicinity of the wall collapses well in wall units (figure 4.28a), the long wavelength part in the outer-flow region tends towards bulk scaling (figure 4.28b). Note that for low Re_τ the scales in the spectra are not sufficiently separated and a substantial fraction of the flow domain is dominated by buffer-layer structures. The footprint of the buffer-layer streaks is visible as the peak of the spectrum which is found around $\lambda_z^+ \approx 1000$ at wall distances around $y^+ \approx 15$. Considering the spanwise pre-multiplied energy spectrum of the streamwise velocity, as depicted in figure 4.29, results in a similar picture. Most energy is accumulated around $\lambda_\varphi^+ \approx 120$ at $y^+ \approx 15$ independently of the Reynolds number when length scales are normalised in wall units (figure 4.29a). The outer plateau of the spectrum further away from the wall tends again to scale in bulk units, as demonstrated by the collapse of the

[16]One-dimensional cuts through the spectra at different wall distances can be found in the appendix (figures E.1 and E.2).

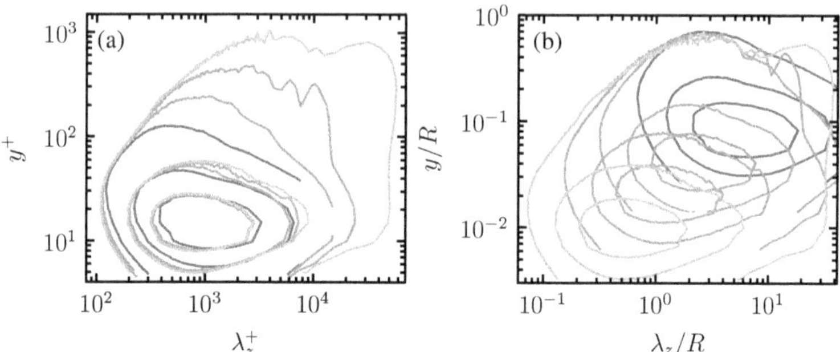

Figure 4.28: One-dimensional streamwise pre-multiplied energy spectra of the streamwise velocity component $\kappa_z\phi_{zz}^+$ as a function of the streamwise wavelength λ_z^+ and the wall distance y^+. Iso-contours values range from 0.4 to 1.6 with an increment of 0.6. (a) lengths in wall units; (b) lengths in bulk units. —, $\mathrm{Re}_\tau = 180$; —, $\mathrm{Re}_\tau = 360$; —, $\mathrm{Re}_\tau = 720$; —, $\mathrm{Re}_\tau = 1500$.

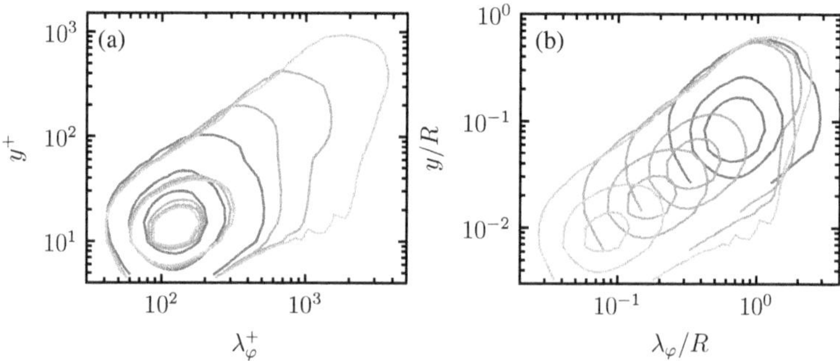

Figure 4.29: One-dimensional spanwise pre-multiplied energy spectra of the streamwise velocity component $\kappa_\varphi\phi_{zz}^+$ as a function of the spanwise wavelength λ_φ^+ and the wall distance y^+. Iso-contour values range from 0.8 to 3.2 with an increment of 1.2. (a) lengths in wall units; (b) lengths in bulk units.—, $\mathrm{Re}_\tau = 180$; , $\mathrm{Re}_\tau = 360$; —, $\mathrm{Re}_\tau = 720$; —, $\mathrm{Re}_\tau = 1500$.

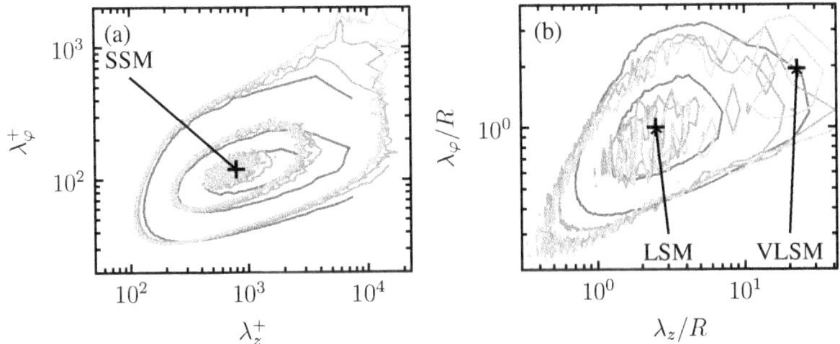

Figure 4.30: Two-dimensional pre-multiplied energy spectra of the streamwise velocity component $\kappa_z \kappa_\varphi \phi_{zz}^+$ as functions of the streamwise wavelength λ_z^+ and the spanwise wavelength λ_φ^+ at two wall distances. (a) $y^+ = 15$, iso-contour values range from 0.1 to 1.0 with an increment of 0.3, wavelengths in wall units; (b) $y/R \approx 0.4$, iso-contour values range from 0.12 to 0.36 with an increment of 0.12, wavelengths in bulk units. —, $\mathrm{Re}_\tau = 180$; —, $\mathrm{Re}_\tau = 360$; —, $\mathrm{Re}_\tau = 720$; —, $\mathrm{Re}_\tau = 1500$.

different Reynolds number spectra at large wavelengths away from the wall in figure 4.29(b). In summary, the footprint of the most energetic coherent structures, namely buffer-layer streaks, is found around $y^+ = 15$ at wavelength of $\lambda_z^+ \approx 1000$ and $\lambda_\varphi^+ \approx 120$. Moreover energetic scales reach lengths up to $\lambda_z/R \approx 30$ and width up to $\lambda_\varphi/R \approx 2$. However, one-dimensional energy spectra in one homogeneous direction are obtained by integration over the other homogeneous direction. Thus, the one-dimensional energy spectrum contains only integral information. Therefore, to obtain the full information about the distribution of energy in the two-dimensional wave number space, two-dimensional pre-multiplied energy spectra are computed and presented at prominent wall distances in figure 4.30. At a wall distance of $y^+ = 15$, presented in figure 4.30(a), the spectra scale well in wall units around the characteristic peak related to buffer-layer streaks. Large wavelengths in stream- and spanwise direction, however, become more energetic as the Reynolds number increases. These wavelengths are assumed to be the footprints of outer-flow VLSMs, which are visible in the large Reynolds number two-dimensional pre-multiplied energy spectra at the wall distance $y/R \approx 0.4$ in figure 4.30(b). At this distance from the wall, the shape of the spectrum changes from a single peak structure for low Reynolds numbers ($\mathrm{Re}_\tau \leq 360$) to a dual peak structure for higher Reynolds numbers ($\mathrm{Re}_\tau \geq 720$). While the second peak is not fully developed at $\mathrm{Re}_\tau = 720$, it becomes the global energy maximum at the wall distance of $y/R \approx 0.4$ at $\mathrm{Re}_\tau = 1500$ with wavelength of $\lambda_z/R \approx 23$ and $\lambda_\varphi/R \approx 2$. This second peak represents

VLSMs, whereas the one around $\lambda_z/R \approx 2.53$ and $\lambda_\varphi/R \approx 1$ at $y/R \approx 0.4$ represents LSMs. While the spectral locations of the latter peaks scale in bulk units, the location of the global spectral energy peak in the vicinity of the wall ($y^+ \approx 15$), which represents SSMs such as buffer-layer streaks, scales in wall units.

An estimate of the average spatial extension of coherent structures both near the wall and in the bulk region in all three dimensions, can be extracted from three-dimensional two-point velocity correlations of the streamwise velocity fluctuation are computed as follows:

$$\rho_{zz}(\Delta z, \Delta\varphi, \Delta r, r_0) = \frac{\langle u_z'(z_0, \varphi_0, r_0) u_z'(z_0 + \Delta z, \varphi_0 + \Delta\varphi, r_0 + \Delta r)\rangle}{u_{z,rms}'^2(r_0)}, \tag{4.28}$$

where r_0 is the radial reference point. Iso-contours of the velocity correlations for the Reynolds numbers considered with a reference point at $y_0/R = 0.4$ ($r_0/R = 0.6$) are presented in figure 4.31. At all Reynolds numbers, the streamwise elongated velocity structures are accompanied by azimuthally adjacent negative correlated structures. Moreover, the iso-contours are attached to the wall at any Reynolds number. As discussed in section 4.1 by means of instantaneous flow field realisations, the LSMs at $y_0/R = 0.4$ are strongly correlated with the SSMs at the wall for $\mathrm{Re}_\tau = 180$, which is indicated by the lengthy wall-penetrating "tails" of the structure visible in figure 4.31(a). Note that at such low Reynolds number SSMs and LSMs are effectively of comparable size. As the near-wall SSMs scale in wall units, their spatial dimension decreases in outer units when increasing the Reynolds number. Thus, I assume a vanishing influence of near-wall SSMs on the outer-flow velocity correlation with increasing Reynolds number (figure 4.31b,c,d). Only for $\mathrm{Re}_\tau = 1500$ the streamwise length of the velocity correlation iso-contour is in the regime of VLSMs. Comparing the velocity correlations between different Reynolds numbers more quantitatively requires two-dimensional cuts through the three-dimensional velocity correlation, as shown in figure 4.32. These cuts correspond to the two-dimensional two-point velocity correlations of the streamwise velocity

$$\rho_{zz}(\Delta z, \Delta r, r_0) = \frac{\langle u_z'(z_0, \varphi_0, r_0) u_z'(z_0 + \Delta z, \varphi_0, r_0 + \Delta r)\rangle}{u_{z,rms}'^2(r_0)}, \tag{4.29}$$

$$\rho_{zz}(\Delta z, \Delta\varphi, r_0) = \frac{\langle u_z'(z_0, \varphi_0, r_0) u_z'(z_0 + \Delta z, \varphi_0 + \Delta\varphi, r_0)\rangle}{u_{z,rms}'^2(r_0)}, \tag{4.30}$$

$$\rho_{zz}(\Delta\varphi, \Delta r, r_0) = \frac{\langle u_z'(z_0, \varphi_0, r_0) u_z'(z_0, \varphi_0 + \Delta\varphi, r_0 + \Delta r)\rangle}{u_{z,rms}'^2(r_0)}. \tag{4.31}$$

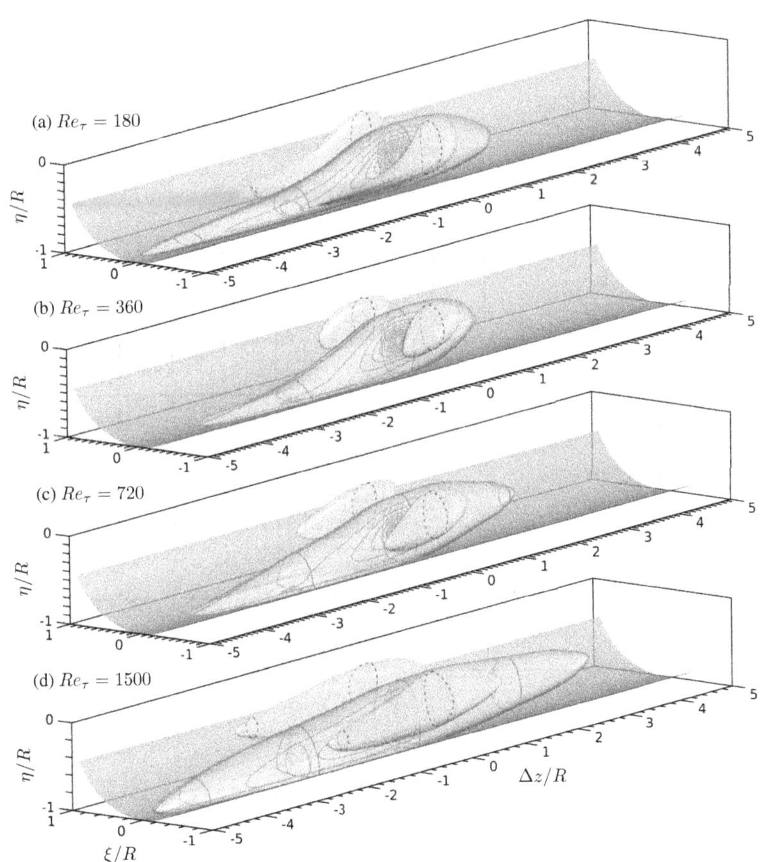

Figure 4.31: Iso-surfaces and iso-contours of the three-dimensional two-point velocity correlation of the streamwise velocity component $\rho_{zz}(\Delta z, \Delta\varphi, \Delta r, r_0)$ as a function of the axial, azimuthal and radial separation lengths. Reference point of the correlation at $y_0/R = 0.4$. Cartesian cross-sectional coordinates $\xi = (r_0 + \Delta r)\sin(\Delta\varphi)$, $\eta = (r_0 + \Delta r)\cos(\Delta\varphi)$. (a) P180; (b) P360; (c) P720; (d) P1500. The orange (cyan) iso-surfaces exhibit values of $+(-)0.1$. Iso-contours values range from $+(-)0.1$ to $+(-)1$, increment of 0.1.

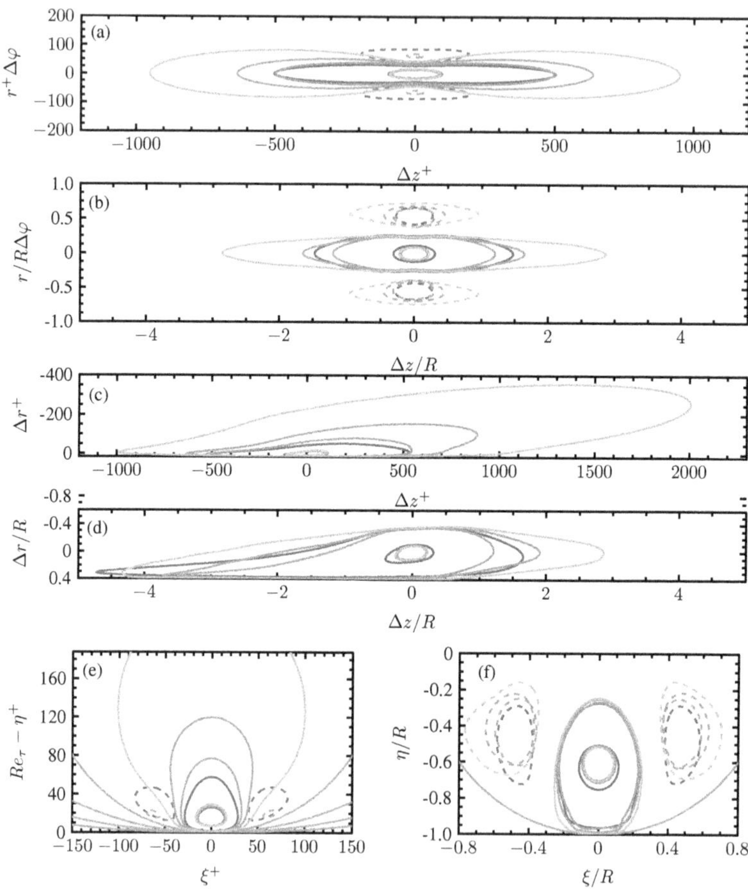

Figure 4.32: Two-dimensional velocity correlations of the streamwise velocity ρ_{zz} at different reference points. (a,b) $\rho_{zz}(\Delta z, \Delta\varphi, r_0)$. (c,d) $\rho_{zz}(\Delta z, \Delta r, r_0)$. (e,f) $\rho_{zz}(\Delta\varphi, \Delta r, r_0)$, cross-sectional coordinates as in figure 4.31. Solid (dashed) iso-contours indicate values of $+(-)0.12, 0.62$. (a,c,e) $y_0^+ = 15$. (b,d,f) $y_0/R = 0.4$. —, P180; —, P360; —, P720; —, P1500.

The correlations are computed involving reference points in the buffer layer ($y_0^+ = 15$, figure 4.32a,c,e) as well as in the bulk flow region ($y_0/R = 0.4$, figure 4.32b,d,f). The values of the contour lines are $\rho_{zz} = 0.62$ and $\rho_{zz} = 0.12$. The latter value is chosen to indicate the same streamwise extension of wall layer streaks as the small-scale peak in the pre-multiplied energy spectra (figure 4.28a and figure 4.30a). Estimating the extension of wall-layer streaks for $Re_\tau \leq 360$ via the $\rho = 0.12$ correlation iso-contour with $y_0^+ = 15$ in the $z\varphi$-plane (figure 4.32a), the zr-plane (figure 4.32c) and the $r\varphi$-plane (figure 4.32e) results in

$$l^+_{z,streak} \times l^+_{\varphi,streak} \times l^+_{r,streak} \approx 1000 \times 50 \times 50, \tag{4.32}$$

which is in good agreement with the literature (Blackwelder and Eckelmann, 1979; Kim et al., 1987; Wagner and Friedrich, 1998). For higher Reynolds numbers ($Re_\tau \geq 720$), the wall-layer streaks correlate with larger regions, a consequence of the streak clustering due to the outer-flow large-scale influence (figure 4.4). Hence, the streamwise extension of velocity streaks is overestimated by the velocity correlation iso-contours. Taking into account pre-multiplied energy spectra (figure 4.30), wall-layer streaks clearly scale in wall units and outer-flow LSMs in bulk units. Regarding the latter structures, figure 4.32(b,d,e) shows a spanwise extension of $\lambda_\varphi \approx 0.5R$ and a streamwise extension of $\lambda_z \approx 3R$ for $Re_\tau \leq 720$. For $Re_\tau = 1500$ the streamwise extension of the velocity iso-contour is $\lambda_z \approx 7R$ and related to VLSM that appear at such high Re_τ. Summarising, the overall size of coherent structures in the bulk flow reads

$$l_{z,LSM} \times l_{\varphi,LSM} \times l_{r,LSM} \approx 3R \times 0.5R \times 0.8R, \tag{4.33}$$

$$l_{z,VLSM} \times l_{\varphi,VLSM} \times l_{r,VLSM} \approx 7R \times 0.5R \times 0.8R, \tag{4.34}$$

when measured by means of velocity correlation iso-contours, which are adjusted via the peak location in pre-multiplied energy spectra. As the three-dimensional velocity correlations (figure 4.31) and the two-dimensional velocity correlations in the zr-plane (figure 4.32c,d) indicate, LSM and VLSM representations are inclined towards the wall, as are their small-scale counterparts. Sillero et al. (2014) measured the inclination angle of velocity correlation iso-contours in TBLs and plane channel flows by approximating them with ellipses and then taking the angle between the major semi axis of these ellipses and the positive x-axes. They pointed out that the inclination angle strongly depends on the iso-contour value being considered. For plane channel flow at $Re_\tau = 950$, a reference point at $y_0/h = 0.4$, and a iso-contour value of 0.1, Sillero et al. (2014) reported an inclination angle of approximately $\alpha \approx 3°$ for the streamwise velocity correlation related to VLSMs. Computing the inclination angle as the angle between the slope that connects the maximum and minimum streamwise coordinate of an iso-contour and the Δz axis results in $\alpha = 0.85°$ for the correlation iso-contour representing velocity streaks ($\mathrm{Re}_\tau = 180$, $y_0^+ = 15$, $\rho_{zz} = 0.12$, figure 4.32c) and $\alpha = 3°$ for the VLSM representation ($\mathrm{Re}_\tau = 1500$, $y_0/R = 0.4$, $\rho_{zz} = 0.12$, figure 4.32d), the latter matching the angle reported by Sillero et al. (2014) well[17].

4.4 Coherent Structure Interaction

As discussed via instantaneous flow field realisations in section 4.1, the outer-flow structure of streamwise velocity fluctuations reflects in the near-wall region. Hoyas and Jiménez (2006) showed that the failure of wall-scaling in turbulent channel flow is caused by large wavelength contributions to the streamwise Reynolds stress. Additionally, Jiménez (2012) demonstrated that the streamwise Reynolds stress profiles collapse in the bulk region of different Reynolds number channel flow, when energies related to structures longer than $6h$ and wider than h are artificially removed. Analogously, a long wavelength cutoff filter is applied to the streamwise fluctuating velocity field of turbulent pipe flow in stream- and spanwise direction. The decomposition of the streamwise fluctuating velocity thus reads

$$u'_z = u'_{zS} + u'_{zL}, \tag{4.35}$$

where u'_{zS} contains scales smaller and u'_{zL} scales larger than the filter length. Figure 4.33 shows the same instantaneous flow field realisation as depicted in figures 4.4(d) and 4.5(d) for $\mathrm{Re}_\tau = 1500$ before (u'_z, figure 4.33a) and after the filtering operation (u'_{zS}, figure 4.33b). Obviously, the very long modes are removed from the flow field both near the wall (figure 4.33 left) and in the bulk region (figure 4.33 right). In the vicinity of the wall the filtered field exhibits regularly alternating low- and high-speed streaks typical for low Reynolds number flow (cf. figure 4.4a). Away from the wall, most of the energy, originally carried by VLSM, is filtered away.

After removing VLSMs by means of a sharp spectral cutoff filter with constant filter length in either wall or bulk units, streamwise Reynolds stresses and skewness are computed. Long-wavelength filtered streamwise Reynolds stress profiles $\langle u'_{zS} u'_{zS} \rangle$ are shown in figure 4.34 for the different Reynolds numbers. The filter lengths are

[17] Note that the current data appears to be more converged than the one reported by Sillero et al. (2014). Hence, I prefer to deduce the inclination angle directly from velocity correlation iso-contours without fitting an ellipse to the data.

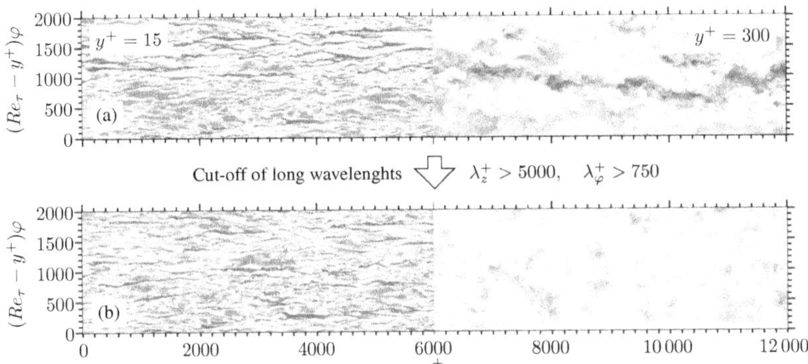

Figure 4.33: Iso-contours of streamwise velocity fluctuations $u_z'^+$ for $\mathrm{Re}_\tau = 1500$. (a) before, (b) after spatial filtering. Left column plane at $y^+ = 15$, right column at $y^+ = 300$. Colour map as in figure 4.4.

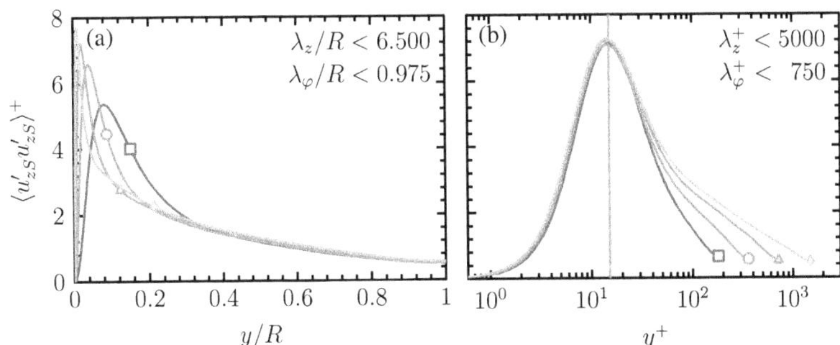

Figure 4.34: Streamwise Reynolds stress computed from a long-wavelength filtered velocity signal. (a) Constant filter length in bulk units, $\lambda_z < 6.5R$, $\lambda_\varphi < 0.975R$. (b) Constant filter lengths in wall units, $\lambda_z^+ < 5000$, $\lambda_\varphi^+ = 750$. -□-, P180; -○-, P360; -△-, P720; -◇-, P1500.

kept constant in bulk units (figure 4.34a) and in wall units (figure 4.34b), respectively. Unlike the unfiltered streamwise Reynolds stress profiles shown in figure 4.16, the filtered profiles scale in wall units both in the near-wall (figure 4.34b) and the bulk region (figure 4.34a). This implies that the scaling failure of streamwise Reynolds stresses is caused by VLSMs, whose energy contribution increases with the Reynolds number.

Expression (4.35) is substituted into equation (1.21) for the streamwise velocity skewness, which yields

$$S(u_z) = \frac{1}{\langle u_z'^2 \rangle^{3/2}} \left(\langle u_{zL}'^3 \rangle + 3\langle u_{zL}'^2 u_{zS}' \rangle + 3\langle u_{zL}' u_{zS}'^2 \rangle + \langle u_{zS}'^3 \rangle \right). \tag{4.36}$$

Using a streamwise scale-decomposition with a filter length of $\lambda_z^+ = 7000$, Mathis et al. (2011) analysed expression (4.36) for experimental TBL. They found that the Reynolds number dependency of the streamwise velocity skewness is caused by the correlation term $3\langle u_{zL}' u_{zS}'^2 \rangle$ only. This term is related to the amplitude modulation term introduced by Mathis et al. (2009a), and thus, responsible for the modulation of the small scales by the large. For the regime of Reynolds numbers considered here scales are, however, not sufficiently separated to apply a streamwise decomposition only. Consequently, the above-introduced two-dimensional decomposition in both streamwise and spanwise direction is applied to compute the terms on the right hand side of expression (4.36) for turbulent pipe flow. The filter lengths of $\lambda_z^+ = 5000$ in streamwise and $\lambda_\varphi^+ = 750$ in spanwise direction are kept constant in wall units. Figure 4.35(b-e) shows the different contributions to the streamwise skewness. In the vicinity of the wall, the small-scale term $\langle u_{zS}'^3 \rangle / \langle u_z'^2 \rangle^{3/2}$ contributes most to the skewness (figure 4.35e), whereas the correlation term $3\langle u_{zL}'^2 u_{zS}' \rangle / \langle u_z'^2 \rangle^{3/2}$ (figure 4.35c) and the large-scale term $\langle u_{zL}'^3 \rangle / \langle u_z'^2 \rangle^{3/2}$ (figure 4.35b) are close to zero. The correlation term $3\langle u_{zL}' u_{zS}'^2 \rangle / \langle u_z'^2 \rangle^{3/2}$, which is depicted in figure 4.35(d), contributes positively to the streamwise skewness in the vicinity of the wall and negatively in the bulk flow region and depends clearly on the Reynolds number. Mathis et al. (2011) introduced a reconstructed skewness

$$S^*(u_z) = \frac{1}{\langle u_z'^2 \rangle^{3/2}} \left(\langle u_{zL}'^3 \rangle + 3\langle u_{zL}'^2 u_{zS}' \rangle + \langle u_{zS}'^3 \rangle \right), \tag{4.37}$$

where the correlation term $3\langle u_{zL}' u_{zS}'^2 \rangle$ is omitted. For TBL of $2800 \leq \mathrm{Re}_\tau \leq 19000$ the authors of the aforementioned study found the reconstructed skewness to be Reynolds number independent. However, their data lacks any values in the vicinity of the wall ($y^+ < 8$), which is due to the limited near-wall accessibility in experimental

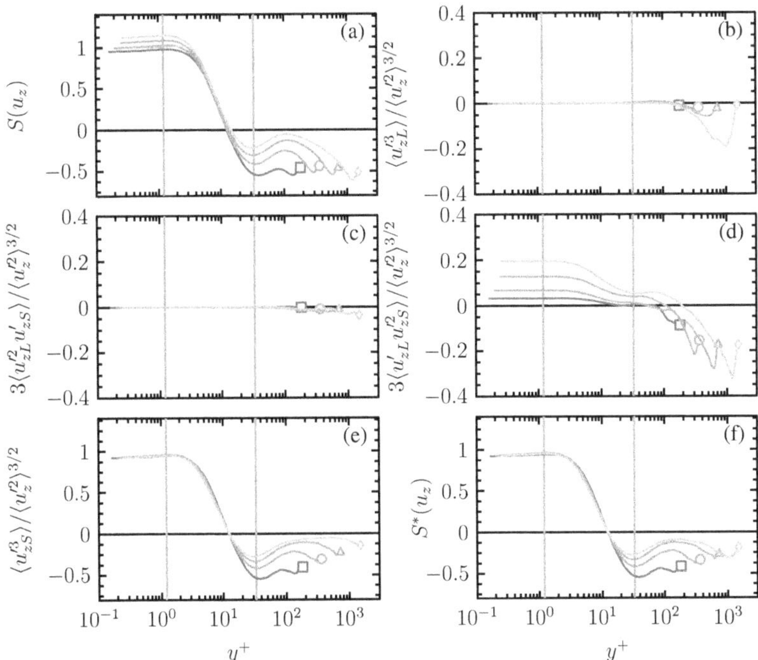

Figure 4.35: (a) Streamwise skewness $S(u_z)$. (b-e) Contributions to the streamwise skewness, based on stream- and spanwise scale decomposition $u'_z = u'_{zL} + u'_{zS}$, filter lengths $\lambda_z^+ = 5000$, $\lambda_\varphi^+ = 750$; (b) $\langle u'^3_{zL}\rangle/\langle u'^2_z\rangle^{3/2}$; (c) $3\langle u'^2_{zL}u'_{zS}\rangle/\langle u'^2_z\rangle^{3/2}$; (d) $3\langle u'_{zL}u'^2_{zS}\rangle/\langle u'^2_z\rangle^{3/2}$; (e) $\langle u'^3_{zS}\rangle/\langle u'^2_z\rangle^{3/2}$. (f) Reconstructed streamwise skewness $S^*(u_z) = (\langle u'^3_{zL}\rangle + 3\langle u'^2_{zL}u'_{zS}\rangle + \langle u'^3_{zS}\rangle)/\langle u'^2_z\rangle^{3/2}$. -□-, P180; -○-, P360; -△-, P720; -◇-, P1500.

investigations. Complementing the latter TBL investigation, the reconstructed skewness of the current pipe flow study, depicted in figure 4.37(f), acknowledges the findings of Mathis et al. (2011) in the near-wall region ($y^+ < 15$), where the reconstructed skewness is Reynolds number independent. Further away from the wall, though, the reconstructed skewness profiles do not scale, which is presumably caused by an insufficient scale separation in the regime of Reynolds numbers considered. Only between the profiles of $\mathrm{Re}_\tau = 720$ and $\mathrm{Re}_\tau = 1500$ the variation is little throughout most of the flow domain. Here, higher Reynolds number profiles are desired to show Reynolds number independence of the reconstructed skewness for turbulent pipe flow.

4.5 Conditional Averaging

The discussion in the preceding sections included the decomposition of coherent structures in positive and negative events — e.g. high- and low-speed streaks. Particularly the correlation between high-speed streaks and (predominately positive) velocity spikes was described by means of low Reynolds number single probe time series (figure 4.2) and instantaneous flow field realisations (figure 4.3). In the upcoming chapter the observations from time series and instantaneous snapshots are underpinned by conditionally averaged statistical quantities. The condition allows the streamwise velocity fluctuation to be only negative or positive. Figure 4.36 displays the conditionally averaged three-dimensional two-point streamwise velocity correlation at a reference point $y_0^+ = 15$ for $\mathrm{Re}_\tau = 180$. The correlations reveal that the head of the low-speed structures is longer than the head of the high-speed structures, whereas the tail is somewhat shorter. Both types of structures feature the characteristic inclination with respect to the wall. A comparison between the high- and the low-speed structure in more detail is presented in figure 4.37, where two-dimensional velocity correlations are shown in planes cutting the three-dimensional correlation at zero separation. From the latter figure quantitative measures can be taken. The head of the low-speed structure is about 100 wall units longer (about ten percent of the full structure length) and its tail is about 50 wall units shorter than the corresponding parts of the high-speed structure. The inclination angle is 0.78° for the low-speed and 1.06° for the high-speed structure (in comparison to 0.85° for the overall correlation, cf. figure 4.32). Computing the correlations at $y_0/R = 0.4$ for $\mathrm{Re}_\tau = 1500$ yields representations of high- and low-speed VLSM, as depicted in figure 4.38. The behaviour of the outer-flow VLSMs is similar to the low Reynolds number near-wall streaks. Low-speed structures (figure 4.38a) feature a longer head and somewhat shorter tail part than high-speed

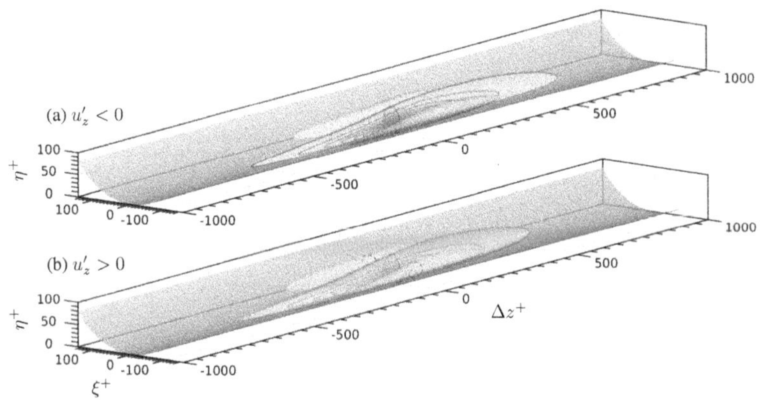

Figure 4.36: Iso-surfaces and iso-contours of the conditionally averaged three-dimensional two-point velocity correlation of the streamwise velocity component $\rho_{zz}(\Delta z, \Delta\varphi, \Delta r, r_0)$ with a reference point at $y_0^+ = 15$ depicted as in figure 4.28. (a) Condition $u'_z < 0$, cyan (orange) iso-surfaces exhibit values of $+(-)0.1$. (b) Condition $u'_z > 0$, orange (cyan) iso-surfaces exhibit values of $+(-)0.1$. $\mathrm{Re}_\tau = 180$.

structures (figure 4.38b). Again, the two-dimensional cuts presented in figure 4.39 allow to measure the difference quantitatively. The head of the low-speed VLSM representation is $1.1R$ (approximately 15 percent) longer, the tail $0.4R$ shorter than the corresponding parts of the high-speed VLSM representation. The inclination angle between both structures varies only little (between 2.9° for the low-speed structure and 3.1° for the high-speed structure). The fact that low-speed structures have longer heads and shorter tails than their high-speed counterpart seems to be related to the wall-normal migration of these structures. Recalling section 1.2.5, low-speed streaks in the near-wall cycle are ejected away from wall, whereas high-speed streaks sweep towards the wall. Thus, the head of low-speed structures penetrates a region with a larger mean convection velocity than the head of low-speed structures. Consequently, the head of the low-speed structure migrates faster in streamwise direction than the head of the high-speed structure resulting in a larger streamwise correlation of the former. The velocity correlations around $y_0/R = 0.4$ (figure 4.39a,b) indicate a similar behaviour for VLSMs. The correlations in the cross-sectional plane (figure 4.39c) reveal the wall-normal organisation of VLSMs of different sign. Negatively correlated regions occur further away from the wall for the correlations of positive streamwise fluctuations than for the ones based on negative streamwise fluctuations, having the reference point

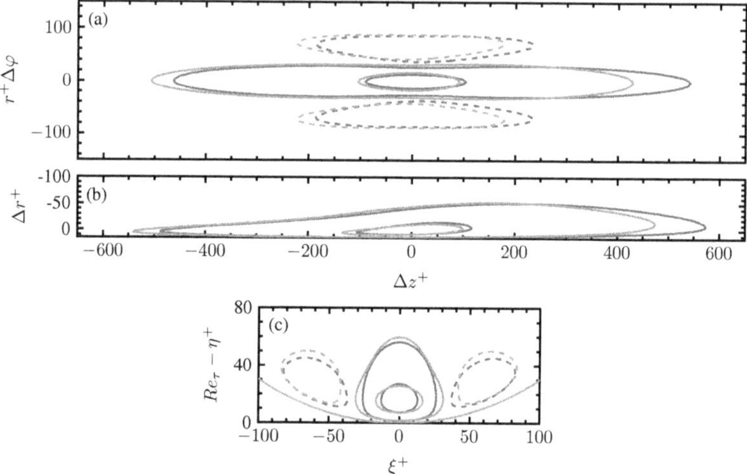

Figure 4.37: Iso-contours of the conditionally averaged two-dimensional velocity correlations of the streamwise velocity ρ_{zz} at different reference points. (a) $\rho_{zz}(\Delta z, \Delta\varphi, r_0)$. (b) $\rho_{zz}(\Delta z, \Delta r, r_0)$. (c) $\rho_{zz}(\Delta\varphi, \Delta r, r_0)$, cross-sectional coordinates as in figure 4.31. Reference point at $y_0^+ = 15$. $\mathrm{Re}_\tau = 180$. Solid (dashed) iso-contours indicate values of $+(-)0.12, 0.62$. —, condition $u_z' < 0$; —, condition $u_z' > 0$.

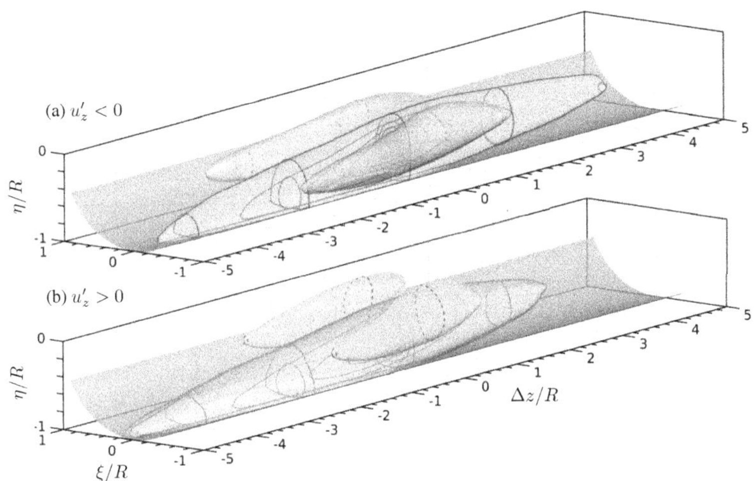

Figure 4.38: Iso-surfaces and iso-contours of the conditionally averaged three-dimensional two-point velocity correlation of the streamwise velocity component $\rho_{zz}(\Delta z, \Delta\varphi, \Delta r, r_0)$ with a reference point at $y_0/R = 0.4$ depicted as in figure 4.28. (a) Condition $u'_z < 0$, cyan (orange) iso-surfaces exhibit values of $+(-)0.1$. (b) Condition $u'_z > 0$, orange (cyan) iso-surfaces exhibit values of $+(-)0.1$. $\mathrm{Re}_\tau = 1500$.

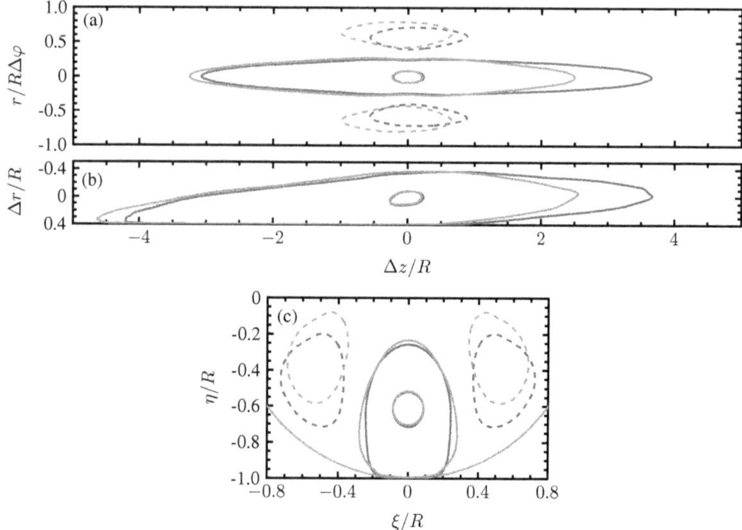

Figure 4.39: Iso-contours of the conditionally averaged two-dimensional velocity correlations of the streamwise velocity ρ_{zz} at different reference points. (a) $\rho_{zz}(\Delta z, \Delta\varphi, r_0)$. (b) $\rho_{zz}(\Delta z, \Delta r, r_0)$. (c) $\rho_{zz}(\Delta\varphi, \Delta r, r_0)$, cross-sectional coordinates as in figure 4.31. Reference point at $y_0/R = 0.4$. $\mathrm{Re}_\tau = 1500$. Solid (dashed) iso-contours indicate values of $+(-)0.12, 0.62$. —, condition $u'_z < 0$; —, condition $u'_z > 0$.

at the same distance from the wall. This indicates that low-speed VLSM are located further away from the wall than azimuthally adjacent high-speed structures. The latter effect is less prominent for the near-wall streaks (figure 4.37c).

As discussed in section 4.4, the logarithmic Reynolds number dependency of near-wall statistics is caused by the interaction of VLSMs with the near-wall cycle. In the following, single contributions from high- and low-speed motions to one-point statistics are investigated by means of conditional averaging. Figure 4.40 displays the streamwise velocity variance (a) and skewness (b) for regions of positive (solid lines) and negative velocity fluctuations (dashed lines). Both near-wall streamwise variance and skewness profiles of the high-speed regions depend significantly stronger on the Reynolds number than the low-speed profiles. This reflects the predominant modulation of small near-wall scales by high-speed outer-flow motions, which is in good agreement with observations of the amplitude modulation in TBL by Mathis et al. (2009a, 2011, 2009b). Furthermore is the high-speed streamwise Reynolds stress peak located closer to the wall ($y^+ \approx 12$) than the low-speed streamwise Reynolds stress peak ($y^+ \approx 18$, figure 4.40a), which reflects that high-speed streaks occur closer to the wall than low-speed streaks. This is in overall agreement with the concept of the near-wall cycle, where quasi-streamwise vortices transport high-speed fluid toward the wall and low-speed fluid away from the wall (cf. section 1.2.5). Also in the outer-flow region, where LSMs and VLSMs are dominant, the low-speed profiles lie above the high-speed ones. The latter matches with the inspection of figure 4.39(c), where low-speed VLSMs are found be located further away from the wall than their high-speed counterparts. In terms of velocity spikes, which are closely linked to streamwise momentum structures (cf. section 4.1.1), the conditionally averaged wall-normal velocity flatness presented in figure 4.41 reveals a similar effect. The latter figure underpins the observations from instantaneous flow fields (cf. figure 4.3), as the flatness value obtained from low-speed streamwise momentum regions is relatively small in the vicinity of the wall. whereas the near-wall flatness value obtained from high-speed streamwise momentum regions is large. In addition, the low-speed profile of the wall-normal flatness exhibits no Reynolds number dependency, while the high-speed profile clearly depends on the Reynolds number near the wall. Since velocity spikes are primarily settled in high-speed regions, the corresponding flatness profiles are subject to the alignment of high-speed streaks into groups, and thus, the spatial redistribution of velocity spikes for high Reynolds numbers. Low-speed regions, on the contrary, carry much less velocity spikes, which results in Reynolds number independent low-speed flatness profiles. In summary, the similar scaling failures of the streamwise Reynolds stress profile, the streamwise

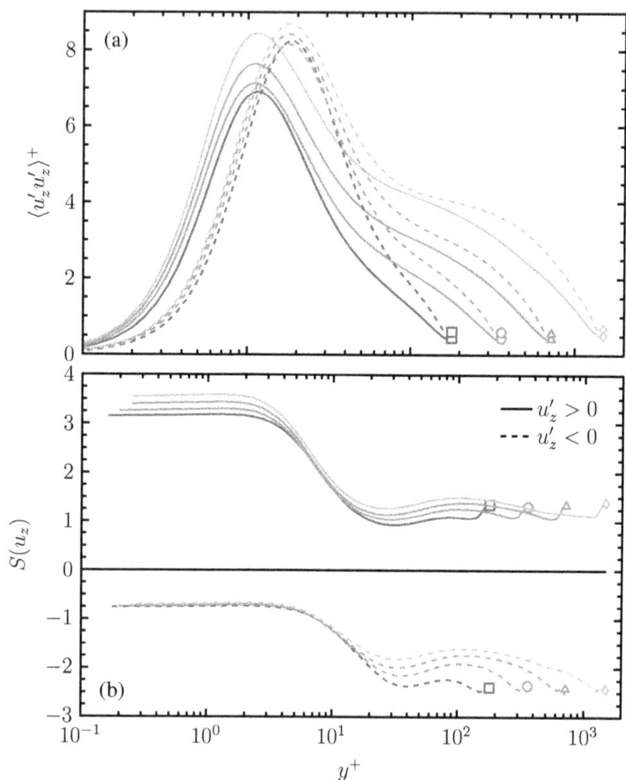

Figure 4.40: Variance $\langle u'_z u'_z \rangle$ (a) and skewness $S(u_z)$ (b) of the streamwise velocity distribution. Conditionally averaged for regions of positive (solid lines) and negative (dashed lines) streamwise velocity fluctuations. -□-, P180; -○-, P360; -△-, P720; -◇-, P1500.

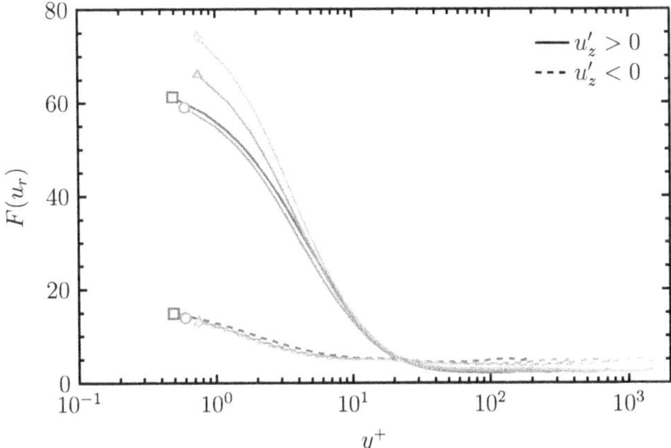

Figure 4.41: Wall-normal flatness component $F(u_r)$, conditionally averaged based on the streamwise velocity fluctuation. —, $F(u_r)$, where $u'_z > 0$; - - -, $F(u_r)$, where $u'_z < 0$. -□-, P180; -○-, P360; -△-, P720; -◇-, P1500.

skewness profile, and the wall-normal flatness profile are presumably caused by the interaction of VLSMs with the near-wall region. Although VLSMs appear in the streamwise velocity component only, they implicitly affect the wall-normal velocity component via the correlation between velocity spikes and high-speed streaks.

4.6 Concluding Remarks

Turbulent pipe flow DNSs of different Reynolds numbers and different wall-normal grid refinement levels were carried out and investigated in terms of instantaneous flow-field realisations, as well as one- and two-point statistics. After the initial characterisation of coherent structures in turbulent pipe flow via instantaneous flow field realisations (section 4.1), the convergence and scaling behaviour of turbulence one-point statistics was elaborated on (section 4.2). Regarding the statistical convergence, spatio-temporal averaging intervals required for reaching a relative error of less than one percent with respect to the true mean were estimated a priori. The estimated intervals were then compared to the effective intervals needed to fulfil the one percent criterion. While low-order statistical quantities, such as the velocity variance, are already sufficiently converged after being averaged over $\Delta\tau^+ = 10^{10}$, higher order statistics need to be integrated much longer in time, which yields $\Delta\tau^+ \approx 2.6 \cdot 10^{11}$ for the wall-normal velocity skewness and $\Delta\tau^+ \approx 6.2 \cdot 10^{11}$ for the wall-normal velocity flatness. Here, not only the order of the statistical moment, but also the intermittency of underlying coherent motions, has an impact on the required convergence interval. In particular, the slow convergence in the wall-normal flatness component is caused by velocity spikes appearing only rarely in space and time.

Regarding the convergence of turbulence statistics with respect to the computational grid, the wall-normal velocity component in the vicinity of the wall is again of particular interest. Coming from a steep gradient, the wall-normal flatness profile asymptotically approaches a plateau towards the wall. While the plateau is not fully recovered from the resolutions used in the study, the flatness profiles are independent of the wall-normal grid spacing up to $\Delta r^+_{min} = 0.5$. Therefore, the latter value was further used for the different Re_τ simulation cases in the study of the dependency of turbulence statistics on the Reynolds number.

The analysis of the scaling behaviour of turbulence one-point statistics manifests logarithmic dependencies of the streamwise velocity variance (for $\mathrm{Re}_\tau \geq 360$), the streamwise velocity skewness, and the wall-normal velocity flatness (for $\mathrm{Re}_\tau \geq 360$) on the Reynolds number in the vicinity of the wall. Away from the wall, however, all flatness components, as well as spanwise and wall-normal velocity variance and skewness components, scale well over the range of the Re_τ considered. The streamwise velocity variance and skewness, in contrast, scale only up to $\mathrm{Re}_\tau = 720$. For $\mathrm{Re}_\tau = 1500$, their profiles show a sudden increase ($\langle u'_z u'_z \rangle$) or decrease ($S(u_z)$) in value, which is presumably a result of VLSMs becoming increasingly energetic at this Re_τ in the outer flow.

In terms of vorticity fluctuations, the contribution from outer-flow VLSMs reflects in the wall-parallel vorticity fluctuation intensities close to the wall ($y^+ \lesssim 10$). The latter depend on the Reynolds number after being normalised in wall units. The same is true for TKE dissipation and viscous diffusion in the same region. Both observed scaling failures agree with the turbulent plane channel flow study by Hoyas and Jiménez (2008), who concluded that the contribution of VLSMs to the enstrophy is restricted to a much thinner region close to the wall than their contribution to energy. In the outer-flow region of turbulent pipe flow the vorticity fluctuation intensities in wall units and the TKE budget terms in wall units scale reasonably well when multiplied by $\sqrt{\mathrm{Re}_\tau}$ or Re_τ, respectively, which is in good agreement with turbulent plane channel flow.

The influence of coherent structures, such as VLSMs, on the scaling behaviour of turbulence statistics was then further discussed in sections 4.3–4.5. After the scaling of coherent structures in turbulent pipe flow was explained via velocity correlations and pre-multiplied energy spectra (section 4.3), their contribution on single point statistics was evaluated by means of spatially decomposed velocity statistics and conditional averages (section 4.4 and 4.5). The analysis demonstrated that the scaling failure in the streamwise Reynolds stress and velocity skewness is caused by large-scale outer-flow motions interacting with the near-wall region, which is consistent with the literature (Hoyas and Jiménez, 2006; Mathis et al., 2011). Consequently, the deletion of large scales from velocity fields leads to Reynolds number independent profiles of the streamwise Reynolds stress. For the streamwise skewness, on the other hand, a Reynolds number independent profile is only achieved in the close vicinity of the wall ($y^+ \leq 15$), when omitting the correlation term $\langle u'_{zL} u'^2_{zS} \rangle$. The fact that the reconstructed streamwise skewness profile depends on the Reynolds number further away from the wall is presumably the effect of an insufficient scale separation in the Reynolds number regime considered. The behaviour of the wall-normal flatness profile in the vicinity of the wall is closely linked to the appearance of velocity spikes, whereas the streamwise Reynolds stress profile reflects the existence of buffer-layer streaks. Since velocity spikes are strongly correlated with high-speed streaks, the interaction of outer-flow VLSMs with the buffer-layer streaks also affects velocity spikes implicitly. As a consequence, the scaling failure of the wall-normal flatness profile is similar to the one in the streamwise Reynolds stress profile. Moreover, conditional averages illustrate that the high wall-normal flatness level at the wall, as well as the Reynolds number dependency of the wall-normal flatness, is related to high-speed regions, only. The latter is underpinned by instantaneous observations that velocity spikes are correlated with high-speed streaks. In summary, VLSMs appearing in the outer-flow region at high-Reynolds-number pipe flow penetrate into the near-wall region, and therefore, interact with the near-wall cycle, which causes a scaling failure of near-wall statistics. Particularly the spatial distribution of velocity streaks changes for high Re_τ, where the latter align in packages that correspond to outer-flow VLSMs. Wall-normal velocity spikes, which are correlated with high-speed streaks realign accordingly, resulting in a scaling failure of the wall-normal flatness profile near the wall.

For the lowest Reynolds number considered ($\mathrm{Re}_\tau = 180$), the value of the wall-normal flatness for turbulent plane channel flow reported by Lenaers et al. (2012) $F(v) \approx 23.3$ at $y^+ = 1$ is much lower than the value for turbulent pipe flow $F(u_r) \approx 30$. For higher Re_τ, on the other hand, the discrepancy is much less. This behaviour is most likely caused by the interplay between velocity spikes and the different flow geometry. Since the latter motions scale in viscous units, they only "see" the curvature of the pipe for low Re_τ. For high Reynolds numbers, the pipe curvature is much less in wall units and the geometry, thus, more channel-alike.

Chapter 5

Scale-by-scale Energy Budget

This chapter unveils contributions from different scales to the turbulent kinetic energy via analysis of the scale-by-scale energy budget for $\mathrm{Re}_\tau = 1500$. As mentioned in section 1.2.6, the decomposition of the turbulence structure into different scales and the determination of the amount of energy that is transferred between different scales is an effective tool to understand the underlying physics. Hence, I analyse the scale-energy budget for scales that are related to the dominant energetic coherent motions in turbulent pipe flow, which are described in chapter 4, after the filtering method and the relevant equations are introduced. Parts of the analysis have been published in Bauer et al. (2019) and Bauer and Wagner (2019a).

5.1 Filtering

The analysis of scale-dependent energy budgets involves the decomposition of the velocity signal via spatial filtering as follows

$$u_i = \overline{u_i} + \widetilde{u}_i, \tag{5.1}$$

where the overbar denotes a filtered quantity and the tilde its residual. The filtering operation is defined as

$$\overline{u_i}(\mathbf{x},t) = \int G(\mathbf{r},\mathbf{x})u_i(\mathbf{x}-\mathbf{r},t)\,\mathrm{d}\mathbf{r}, \tag{5.2}$$

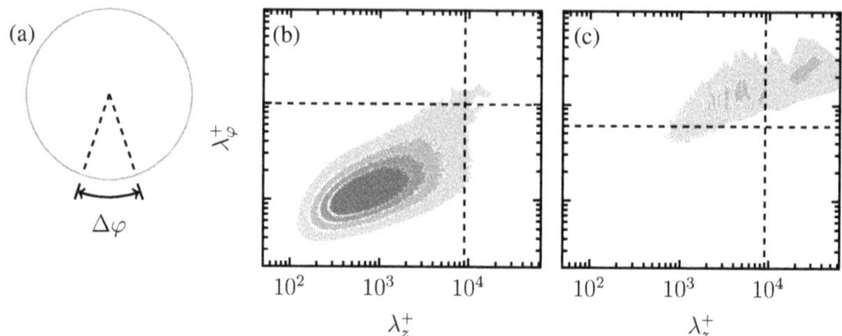

Figure 5.1: (a) Spanwise filter in the pipe cross section. (b,c) Two-dimensional pre-multiplied energy spectra of the streamwise velocity component $\kappa_z\kappa_\varphi\phi_{zz}$ as functions of the streamwise wavelength λ_z and the spanwise wavelength λ_φ at $y^+ = 15$ (b) and $y/R = 0.4$ (c). All units are normalised by wall units. $\kappa_c = \pi/\lambda$. Iso-surfaces denote values ranging from 0.12 (light red) to 0.48 (dark red) with an increment of 0.12.

with $G(r)$ the filter function. In the following, G particularly denotes a sharp spectral cut-off filter[18] defined as

$$G(\kappa) = \begin{cases} 0 & \text{if } \kappa \leq \kappa_c \\ 1 & \text{if } \kappa > \kappa_c \end{cases} \tag{5.3}$$

in spectral space with a cut-off wave number $\kappa_c = \pi/\lambda$. Paying tribute to the flow geometry, the filter is applied in the two homogeneous directions (z,φ), whereas no filter is applied in the radial direction (r). To connect the scale-energy budgets with particular energetic coherent structures, the filter lengths are chosen to divide the range of scales between the energetic modes. Since these structures are longer than they are wide, see e.g. section 4.3.2, anisotropic filtering is applied. To divide the range of scales into SSMs on one side and LSMs as well as VLSMs on the other side, Cho et al. (2018) and Kawata and Alfredsson (2018) used only one-dimensional filters in spanwise direction for their channel flow analysis. As LSMs and VLSMs involve similar spanwise wavelengths additional filtering in streamwise direction, where the latter motions separate, is required for the current analysis. Figure 5.1 shows the filter geometry together with pre-multiplied energy spectra of the streamwise velocity. As indicated in figure 5.1(a), the spanwise filter length $\Delta\varphi$ is kept constant in radian measure, whereas the streamwise filter length λ_z is fixed in spatial units. Since the

[18]The chosen filter type has several advantages. As a spatially uniform filter, it commutes with integration and differentiation. Moreover, the filter is a projection, and thus, the filtered residual is zero ($\overline{\overline{u}} = \overline{u}$, $\overline{\widetilde{u}} = 0$).

present analysis focuses on the flow region near the wall — i.e. on the buffer layer where SSMs are dominant and on the logarithmic layer where LSMs and VLSMs play a role —, the spanwise filter length at the wall $\lambda_\varphi = R\Delta\varphi$ is given for comparison. In the present study, two different filter lengths are considered. The first, hereinafter denoted as configuration A with filter lengths $\lambda_z^+ = 76$ and $\lambda_\varphi^+ = R^+\Delta\varphi = 42$, is equivalent to the one used in Härtel et al. (1994). As these filter lengths represent the typical grid spacing in LES, the filtered motions are comparable to subgrid-scale structures in LES. Filter configuration A unveils effects associated to viscous SSMs. Via the second filter configuration (B), on the other hand, the behaviour of VLSMs, which are less energetic than the near-wall SSMs, is uncovered. Dividing the range of scales between SSMs and LSMs on the one, and VLSMs on the other hand, filter length are determined via two-dimensional pre-multiplied spectra, as depicted in figure 5.1(b,c). Filter lengths of $\lambda_z^+ = 9000$ and $\lambda_\varphi^+ = 1000$ ($\Delta\varphi = 2/3$) satisfy the condition and are indicated by dashed lines in figure 5.1. In particular, the latter figure, where pre-multiplied energy spectra of the streamwise velocity are shown at two distances from the wall ($y^+ = 15$, figure 5.1b; $y/R = 0.4$ figure 5.1c), illustrates that filter configuration B separates VLSMs (dark shaded areas in the upper right corner of subfigure c) from both LSMs (dark shaded areas in the upper half of subfigure c around $\lambda_z^+ \approx 3000$) and SSMs (dark shaded areas in the lower left corner of subfigure b). Although Hultmark et al. (2012) showed that the energy content of VLSMs increases with the Reynolds number, a Reynolds number as low as $Re_\tau = 1500$ already manifests a clear separation of scales necessary for the analysis of the interaction between VLSMs and smaller scales. Figure 5.1 also corroborates that the results are not affected by the variation of the spanwise filter length in wall units in the region of interest, where energetic turbulent scales are separated sufficiently.

5.2 Derivation of Energy Transport Equations

The application of the filter above decomposes the fluctuating velocity field into two scale components, hereinafter referred to as filtered and subfilter scale. Taking into account both the Reynolds decomposition (1.17) and the filtering operation, the velocity signal is decomposed into three contributions, namely the mean velocity, the filtered fluctuating velocity and the subfilter fluctuating velocity[19]. Hence, the expression for

[19]Since I apply filtering in the homogeneous directions only, statistical means, which are constants at each wall distances are not affected by the filter.

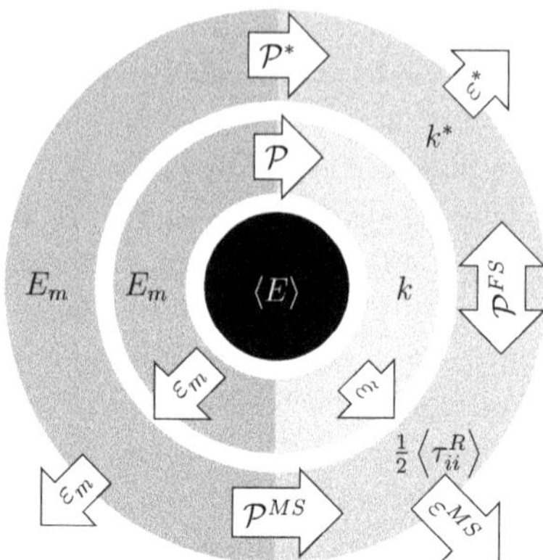

Figure 5.2: Scheme of different contributions to the mean kinetic energy $\langle E \rangle$ based on the velocity decomposition.

the mean kinetic energy (1.56) is evaluated to

$$\langle E \rangle = \underbrace{\frac{1}{2}\langle u_i \rangle \langle u_i \rangle}_{E_m} + \underbrace{\frac{1}{2}\langle \overline{u_i'} \cdot \overline{u_i'} \rangle}_{k^*} + \frac{1}{2}\langle \tau_{ii}^R \rangle, \tag{5.4}$$

where

$$\tau_{ij}^R = \overline{u_i u_j} - \overline{u_i} \cdot \overline{u_j} = \overline{u_i' u_j'} - \overline{u_i'} \cdot \overline{u_j'} \tag{5.5}$$

is the residual stress. The contributions to the mean kinetic energy in equation (5.4) are presented in figure 5.2 together with the flux terms, which are responsible for the energy transfer between the different contributions. These terms are found as source and sink terms in the transport equation of the corresponding quantity.

Transport equations for the turbulent kinetic energy k and the kinetic energy of the mean velocity E_m are already derived in the fundamentals, see equations (1.58) and (1.59). In addition, transport equations for k^* and $1/2\langle \tau_{ii}^R \rangle$ are derived hereinafter. Finally, the streamwise Reynolds stress budget of the filtered velocity field, which forms the basis for the following analysis of energy-containing coherent structures, is derived. First, the non-dimensional momentum equation (1.9b) are stated in a slightly different form

$$\frac{\partial u_j}{\partial t} + \frac{\partial u_i u_j}{\partial x_i} = \frac{1}{\mathrm{Re}_\tau} \frac{\partial^2 u_j}{\partial x_i \partial x_i} - \frac{\partial p}{\partial x_j}. \tag{5.6}$$

Applying the filtering operation (5.2) to both the momentum equation (5.6) and the Reynolds equation (1.57b) results in

$$\frac{\partial \overline{u}_j}{\partial t} + \frac{\partial \overline{u_i u_j}}{\partial x_i} = \frac{1}{\mathrm{Re}_\tau} \frac{\partial^2 \overline{u}_j}{\partial x_i \partial x_i} - \frac{\partial \overline{p}}{\partial x_j}, \tag{5.7}$$

and

$$\frac{\partial \langle \overline{u}_j \rangle}{\partial t} + \frac{\partial \overline{\langle u_i \rangle \langle u_j \rangle}}{\partial x_i} = -\frac{\partial \langle \overline{u'_i u'_j} \rangle}{\partial x_i} + \frac{1}{Re_\tau} \frac{\partial^2 \langle \overline{u}_j \rangle}{\partial x_i \partial x_i} - \frac{\partial \langle \overline{p} \rangle}{\partial x_j}, \tag{5.8}$$

respectively[20]. Subtracting the filtered Reynolds equation (5.8) from the filtered momentum equation (5.7) gives an equation for the filtered fluctuating velocity

$$\begin{aligned}
\frac{\partial(\overline{u}_j - \langle \overline{u}_j \rangle)}{\partial t} + \frac{\partial(\overline{u_i u_j} - \overline{\langle u_i \rangle \langle u_j \rangle})}{\partial x_i} &= \frac{\partial \langle \overline{u'_i u'_j} \rangle}{\partial x_i} + \frac{1}{\mathrm{Re}_\tau} \frac{\partial^2 (\overline{u}_j - \langle \overline{u}_j \rangle)}{\partial x_i \partial x_i} - \frac{\partial (\overline{p} - \langle \overline{p} \rangle)}{\partial x_j} \\
\Leftrightarrow \frac{\partial \overline{(u_j - \langle u_j \rangle)}}{\partial t} + \frac{\partial \overline{(u_i u_j - \langle u_i \rangle \langle u_j \rangle)}}{\partial x_i} &= \frac{\partial \langle \overline{u'_i u'_j} \rangle}{\partial x_i} + \frac{1}{Re_\tau} \frac{\partial^2 \overline{(u_j - \langle u_j \rangle)}}{\partial x_i \partial x_i} - \frac{\partial \overline{(p - \langle p \rangle)}}{\partial x_j} \\
\Leftrightarrow \frac{\partial \overline{u'_j}}{\partial t} + \frac{\partial (\overline{\langle u_i \rangle u'_j} + \overline{u'_i \langle u_j \rangle} + \overline{u'_i u'_j})}{\partial x_i} &= \frac{\partial \langle \overline{u'_i u'_j} \rangle}{\partial x_i} + \frac{1}{\mathrm{Re}_\tau} \frac{\partial^2 \overline{u'_j}}{\partial x_i \partial x_i} - \frac{\partial \overline{p'}}{\partial x_j}.
\end{aligned} \tag{5.9}$$

Then expression (5.5) is substituted into equation (5.9) and index i is replaced by k, yielding

$$\frac{\partial \overline{u'_j}}{\partial t} + \overline{u'_k} \frac{\partial \overline{u'_j}}{\partial x_k} = \frac{1}{Re_\tau} \frac{\partial^2 \overline{u'_j}}{\partial x_k \partial x_k} + \frac{\partial}{\partial x_k} \left(\langle \tau^R_{kj} \rangle + \langle \overline{u'_k} \cdot \overline{u'_j} \rangle - \overline{\langle u_k \rangle u'_j} - \overline{u'_k \langle u_j \rangle} - \tau^R_{kj} \right) - \frac{\partial \overline{p'}}{\partial x_j}. \tag{5.10}$$

The latter equation (5.10) is now multiplied with $\overline{u'_j}$ to obtain

$$\begin{aligned}
\underbrace{\overline{u'_j} \frac{\partial \overline{u'_j}}{\partial t}}_{T_1} + \underbrace{\overline{u'_j} \cdot \overline{u'_k} \frac{\partial \overline{u'_j}}{\partial x_k}}_{T_2} &= \underbrace{\frac{1}{Re_\tau} \overline{u'_j} \frac{\partial^2 \overline{u'_j}}{\partial x_k \partial x_k}}_{T_3} + \overline{u'_j} \frac{\partial \langle \tau^R_{kj} \rangle}{\partial x_k} + \overline{u'_j} \frac{\partial \langle \overline{u'_k} \cdot \overline{u'_j} \rangle}{\partial x_k} \\
&- \underbrace{\left(\overline{u'_j} \frac{\partial \overline{\langle u_k \rangle u'_j}}{\partial x_k} + \overline{u'_j} \frac{\partial \overline{u'_k \langle u_j \rangle}}{\partial x_k} \right)}_{T_4} - \overline{u'_j} \frac{\partial \overline{p'}}{\partial x_j} - \underbrace{\overline{u'_j} \frac{\partial \tau^R_{kj}}{\partial x_k}}_{T_5},
\end{aligned} \tag{5.11}$$

where term T_1 can be rewritten to

$$\overline{u'_j} \frac{\partial \overline{u'_j}}{\partial t} = \frac{1}{2} \frac{\partial (\overline{u'_j} \cdot \overline{u'_j})}{\partial t}, \tag{5.12}$$

[20] Here I use the commutability between the chosen filter and the averaging operation.

and term T_2 can be transformed into

$$\overline{u'_j} \cdot \overline{u'_k} \frac{\partial(\overline{u'_j})}{\partial x_k} = \frac{1}{2} \overline{u'_k} \frac{\partial(\overline{u'_j} \cdot \overline{u'_j})}{\partial x_k} = \frac{1}{2} \frac{\partial(\overline{u'_j} \cdot \overline{u'_j} \cdot \overline{u'_k})}{\partial x_k}. \tag{5.13}$$

Term T_3 is further decomposed into viscous diffusion and (pseudo-)dissipation

$$\begin{aligned} \frac{1}{Re_\tau} \overline{u'_j} \frac{\partial^2 \overline{u'_j}}{\partial x_k \partial x_k} &= \frac{1}{Re_\tau} \left(\frac{\partial}{\partial x_k} \left(\overline{u'_j} \frac{\partial \overline{u'_j}}{\partial x_k} \right) - \frac{\partial \overline{u'_j}}{\partial x_k} \frac{\partial \overline{u'_j}}{\partial x_k} \right) \\ &= \frac{1}{Re_\tau} \left(\frac{\partial}{\partial x_k} \left(\frac{1}{2} \frac{\partial \overline{u'_j} \cdot \overline{u'_j}}{\partial x_k} \right) - \frac{\partial \overline{u'_j}}{\partial x_k} \frac{\partial \overline{u'_j}}{\partial x_k} \right) \\ &= \frac{1}{2Re_\tau} \frac{\partial^2 (\overline{u'_j} \cdot \overline{u'_j})}{\partial x_k \partial x_k} - \frac{1}{Re_\tau} \frac{\partial \overline{u'_j}}{\partial x_k} \frac{\partial \overline{u'_j}}{\partial x_k}. \end{aligned} \tag{5.14}$$

T_4 is simplified first via application of the divergence-free condition ($\partial u_i / \partial x_i = 0$),

$$\overline{u'_j} \frac{\partial \overline{\langle u_k \rangle u'_j}}{\partial x_k} + \overline{u'_j} \frac{\partial \overline{u'_k \langle u_j \rangle}}{\partial x_k} = \overline{u'_j} \overline{\left(\langle u_k \rangle \frac{\partial u'_j}{\partial x_k} \right)} + \overline{u'_j} \overline{\left(u'_k \frac{\partial \langle u_j \rangle}{\partial x_k} \right)}, \tag{5.15}$$

then further using the fact that the mean velocity is invariant to the filtering operation ($\overline{\langle u_i \rangle} = \langle u_i \rangle$),

$$\overline{u'_j} \frac{\partial \overline{\langle u_k \rangle u'_j}}{\partial x_k} + \overline{u'_j} \frac{\partial \overline{u'_k \langle u_j \rangle}}{\partial x_k} = \overline{u'_j} \langle u_k \rangle \frac{\partial \overline{u'_j}}{\partial x_k} + \overline{u'_j} \cdot \overline{u'_k} \frac{\partial \langle u_j \rangle}{\partial x_k}. \tag{5.16}$$

Finally, T_5 is expanded to

$$\overline{u'_j} \frac{\partial \tau^R_{kj}}{\partial x_k} = \frac{\partial(\overline{u'_j} \tau^R_{kj})}{\partial x_k} - \tau^R_{kj} \frac{\partial \overline{u'_j}}{\partial x_k}. \tag{5.17}$$

Substituting the expressions T_1 to T_5 back into equation (5.11) gives

$$\begin{aligned} & \frac{1}{2} \frac{\partial(\overline{u'_j} \cdot \overline{u'_j})}{\partial t} + \frac{1}{2} \frac{\partial(\overline{u'_j} \cdot \overline{u'_j} \cdot \overline{u'_k})}{\partial x_k} = \frac{1}{2Re_\tau} \frac{\partial^2 (\overline{u'_j} \cdot \overline{u'_j})}{\partial x_k \partial x_k} - \frac{1}{Re_\tau} \frac{\partial \overline{u'_j}}{\partial x_k} \frac{\partial \overline{u'_j}}{\partial x_k} + \overline{u'_j} \frac{\partial \langle \tau^R_{kj} \rangle}{\partial x_k} \\ + & \overline{u'_j} \frac{\partial \langle \overline{u'_k} \cdot \overline{u'_j} \rangle}{\partial x_k} - \overline{u'_j} \langle u_k \rangle \frac{\partial \overline{u'_j}}{\partial x_k} - \overline{u'_j} \cdot \overline{u'_k} \frac{\partial \langle u_j \rangle}{\partial x_k} - \overline{u'_j} \frac{\partial \overline{p'}}{\partial x_j} - \frac{\partial(\overline{u'_j} \tau^R_{kj})}{\partial x_k} + \tau^R_{kj} \frac{\partial \overline{u'_j}}{\partial x_k}, \end{aligned} \tag{5.18}$$

which can be reformulated to

$$\frac{\partial(\frac{1}{2}\overline{u_j'}\cdot\overline{u_j'})}{\partial t}+\frac{\partial}{\partial x_k}\left[(\langle u_k\rangle+\overline{u_k'})(\tfrac{1}{2}\overline{u_j'}\cdot\overline{u_j'})+\overline{u_k'}\cdot\overline{p'}-\frac{1}{Re_\tau}\frac{\partial(\frac{1}{2}\overline{u_j'}\cdot\overline{u_j'})}{\partial x_k}+\overline{u_j'}\tau_{kj}^R\right]$$
$$=\overline{u_j'}\frac{\partial\langle\tau_{kj}^R\rangle}{\partial x_k}+\overline{u_j'}\frac{\partial\langle\overline{u_k'}\cdot\overline{u_j'}\rangle}{\partial x_k}-\overline{u_j'}\cdot\overline{u_k'}\frac{\partial\langle u_j\rangle}{\partial x_k}+\tau_{kj}^R\frac{\partial\overline{u_j'}}{\partial x_k}-\frac{1}{Re_\tau}\frac{\partial\overline{u_j'}}{\partial x_k}\frac{\partial\overline{u_j'}}{\partial x_k}, \tag{5.19}$$

and finally

$$\frac{\partial e^*}{\partial t}+\frac{\partial}{\partial x_k}\left[\overline{u_k}e^*+\overline{u_k'}\cdot\overline{p'}-\frac{1}{\mathrm{Re}_\tau}\frac{\partial e^*}{\partial x_k}+\overline{u_j'}\left(\tau_{kj}^R-\langle\tau_{kj}^R\rangle\right)-\overline{u_j'}\langle\overline{u_j'}\cdot\overline{u_k'}\rangle\right]$$
$$=-\langle\overline{u_k'}\cdot\overline{u_j'}\rangle\frac{\partial\overline{u_j'}}{\partial x_k}-\overline{u_j'}\cdot\overline{u_k'}\frac{\partial\langle u_j\rangle}{\partial x_k}+\left(\tau_{kj}^R-\langle\tau_{kj}^R\rangle\right)\frac{\partial\overline{u_j'}}{\partial x_k}-\frac{1}{\mathrm{Re}_\tau}\frac{\partial\overline{u_j'}}{\partial x_k}\frac{\partial\overline{u_j'}}{\partial x_k}, \tag{5.20}$$

where

$$e^*=\frac{1}{2}\overline{u_j'}\cdot\overline{u_j'} \tag{5.21}$$

is the instantaneous kinetic energy of the filtered fluctuating velocity field. Applying the averaging operator (1.33) on equation (5.19) gives the transport equation of the turbulent kinetic energy of the filtered velocity field k^*,

$$\frac{\partial k^*}{\partial t}+\frac{\partial}{\partial x_k}\left[\langle u_k\rangle k^*+\langle\overline{u_k'}\cdot\overline{p'}\rangle+\frac{1}{2}\langle\overline{u_i'}\cdot\overline{u_i'}\cdot\overline{u_k'}\rangle-\frac{1}{\mathrm{Re}_\tau}\frac{\partial k^*}{\partial x_k}+\langle\overline{u_j'}\tau_{kj}^R\rangle\right]$$
$$=\underbrace{-\langle\overline{u_j'}\cdot\overline{u_k'}\rangle\frac{\partial\langle u_j\rangle}{\partial x_k}}_{\mathcal{P}^*}+\underbrace{\left\langle\tau_{kj}^R\frac{\partial\overline{u_j'}}{\partial x_k}\right\rangle}_{\mathcal{P}^{FS}}-\underbrace{\frac{1}{\mathrm{Re}_\tau}\left\langle\frac{\partial\overline{u_j'}}{\partial x_k}\frac{\partial\overline{u_j'}}{\partial x_k}\right\rangle}_{\varepsilon^*}. \tag{5.22}$$

Comparing the transport equation for k^* with the one for k (cf. equation 1.58) reveals the appearance of two additional terms, which are highlighted by red colour in equation (5.22). The term on the left-hand side is an additional transport term in physical space, which acts on the filtered velocity field via interaction with subfilter scales. Consequently, I refer to this term as subfilter-scale diffusion. The second additional term appearing on the right-hand-side of the equation is an additional source or sink term and represents the transport of energy from or towards scales smaller than the filter scale. In the context of LES, it is often referred to as subgrid-scale dissipation. However, since the term is not necessarily negative, as sketched in section 1.2.6, I denote it as the inter-scale energy flux.

Evaluating equations (5.11) to (5.20) after the unfiltered Reynolds equation (1.57b) has been subtracted from the unfiltered momentum equation (5.6) results in a transport equation of the kinetic energy of the fluctuating field,

$$
\begin{aligned}
\frac{\partial e}{\partial t} &+ \frac{\partial}{\partial x_k}\left[u_k e + u_k' p' - \frac{1}{\mathrm{Re}_\tau}\frac{\partial e}{\partial x_k} - u_j'\langle u_j' u_k'\rangle\right] \\
&= -\langle u_k' u_j'\rangle \frac{\partial u_j'}{\partial x_k} - u_j' u_k' \frac{\partial \langle u_j\rangle}{\partial x_k} - \frac{1}{\mathrm{Re}_\tau}\frac{\partial u_j'}{\partial x_k}\frac{\partial u_j'}{\partial x_k},
\end{aligned} \tag{5.23}
$$

where

$$
e = \frac{1}{2} u_j' u_j' \tag{5.24}
$$

is the instantaneous kinetic energy of the fluctuating velocity field. Subsequently, the filtering operation (5.2) is applied to the transport equation of e (5.23), before the results is subtracted from the transport equation of k^* (5.22). Finally, averaging over the outcome of the latter yields the transport equation for the residual turbulent kinetic energy $1/2\langle\tau_{ii}^R\rangle$,

$$
\begin{aligned}
&\frac{\partial \frac{1}{2}\langle\tau_{ii}^R\rangle}{\partial t} + \frac{\partial}{\partial x_k}\Big[\langle u_k\rangle \tfrac{1}{2}\langle\tau_{ii}^R\rangle + \langle u_k' p'\rangle - \langle\overline{u_k'}\cdot\overline{p'}\rangle \\
&\qquad + \frac{1}{2}\langle u_i' u_i' u_k'\rangle - \frac{1}{2}\langle\overline{u_i'}\cdot\overline{u_i'}\cdot\overline{u_k'}\rangle - \frac{1}{\mathrm{Re}_\tau}\frac{\partial\frac{1}{2}\langle\tau_{ii}^R\rangle}{\partial x_k} - \langle\overline{u_j'}\tau_{kj}^R\rangle\Big] \\
&= \underbrace{-\left(\langle\overline{u_j'}\cdot\overline{u_k'}\rangle + \langle\tau_{kj}^R\rangle\right)\frac{\partial\langle u_j\rangle}{\partial x_k}}_{\mathcal{P}^{MS}} - \underbrace{\left\langle \tau_{kj}^R \frac{\partial \overline{u_j'}}{\partial x_k}\right\rangle}_{\mathcal{P}^{FS}} - \underbrace{\frac{1}{\mathrm{Re}_\tau}\left(\left\langle\frac{\partial u_j'}{\partial x_k}\frac{\partial u_j'}{\partial x_k}\right\rangle - \left\langle\frac{\partial \overline{u_j'}}{\partial x_k}\frac{\partial \overline{u_j'}}{\partial x_k}\right\rangle\right)}_{\varepsilon^{MS}}.
\end{aligned} \tag{5.25}
$$

Obviously, the production of the unfiltered field is the sum of the contributions from the filtered field and the residual field, i.e.

$$
\mathcal{P} = \mathcal{P}^* + \mathcal{P}^{MS}, \tag{5.26}
$$

while the overall dissipation is compound by the dissipation of the filtered and the residual field, i.e.

$$
\varepsilon = \varepsilon^* + \varepsilon^{MS}, \tag{5.27}
$$

see also figure 5.2. In analogy to the production term $\mathcal{P}$, which exchanges kinetic energy between the mean and the fluctuating field, the inter-scale energy flux term $\mathcal{P}^{FS}$ is responsible for the kinetic energy exchange between the fluctuating filtered field and the residual field. Unlike the production term $\mathcal{P}$, the inter-scale energy flux term $\mathcal{P}^{FS}$ can be a source or a sink in both transport equations (5.22) and (5.25).

Additionally, in analogy to the Reynolds stress budget equation — see Pope (2000), page 315, equation (7.178) — a transport equation for the Reynolds stresses of the filtered velocity field can be derived. Therefore, I multiply equation (5.10) for component i and j with $\overline{u'_j}$ and $\overline{u'_i}$, respectively, leading to

$$\begin{aligned}
\overline{u'_i}\frac{\partial \overline{u'_j}}{\partial t} + \overline{u'_i} \cdot \overline{u'_k}\frac{\partial \overline{u'_j}}{\partial x_k} &= \frac{1}{\mathrm{Re}_\tau}\overline{u'_i}\frac{\partial^2 \overline{u'_j}}{\partial x_k \partial x_k} \\
&+ \overline{u'_i}\frac{\partial}{\partial x_k}\left(\langle u'_k u'_j\rangle - \langle u_k\rangle \overline{u'_j} - \overline{u'_k}\langle u_j\rangle - \tau^R_{kj}\right) - \overline{u'_i}\frac{\partial \overline{p'}}{\partial x_j}, \quad (5.28)\\
\overline{u'_j}\frac{\partial \overline{u'_i}}{\partial t} + \overline{u'_j} \cdot \overline{u'_k}\frac{\partial \overline{u'_i}}{\partial x_k} &= \frac{1}{\mathrm{Re}_\tau}\overline{u'_j}\frac{\partial^2 \overline{u'_i}}{\partial x_k \partial x_k} \\
&+ \overline{u'_j}\frac{\partial}{\partial x_k}\left(\langle u'_k u'_i\rangle - \langle u_k\rangle \overline{u'_i} - \overline{u'_k}\langle u_i\rangle - \tau^R_{ki}\right) - \overline{u'_j}\frac{\partial \overline{p'}}{\partial x_j}. \quad (5.29)
\end{aligned}$$

Next, the summation of (5.28) and (5.29) results in

$$\begin{aligned}
\frac{\partial(\overline{u'_i}\cdot\overline{u'_j})}{\partial t} &+ \underbrace{\overline{u'_k}\frac{\partial(\overline{u'_i}\cdot\overline{u'_j})}{\partial x_k}}_{T'_1} = \underbrace{\frac{1}{\mathrm{Re}_\tau}\left(\overline{u'_i}\frac{\partial^2 \overline{u'_j}}{\partial x_k\partial x_k} + \overline{u'_j}\frac{\partial^2 \overline{u'_i}}{\partial x_k\partial x_k}\right)}_{T'_2)} \\
&+ \overline{u'_i}\frac{\partial\langle u'_k u'_j\rangle}{\partial x_k} + \overline{u'_j}\frac{\partial\langle u'_k u'_i\rangle}{\partial x_k} \\
&- \underbrace{\overline{u'_i}\frac{\partial\langle u_k\rangle\overline{u'_j}}{\partial x_k} - \overline{u'_j}\frac{\partial\langle u_k\rangle\overline{u'_i}}{\partial x_k} - \overline{u'_i}\frac{\partial\overline{u'_k}\langle u_j\rangle}{\partial x_k} - \overline{u'_j}\frac{\partial\overline{u'_k}\langle u_i\rangle}{\partial x_k}}_{T'_3} \\
&- \overline{u'_i}\frac{\partial\tau^R_{kj}}{\partial x_k} - \overline{u'_j}\frac{\partial\tau^R_{ki}}{\partial x_k} - \overline{u'_i}\frac{\partial\overline{p'}}{\partial x_j} - \overline{u'_j}\frac{\partial\overline{p'}}{\partial x_i}. \quad (5.30)
\end{aligned}$$

Using once more the divergence-free condition, term T'_1 is reformulated to

$$\overline{u'_k}\frac{\partial(\overline{u'_i}\cdot\overline{u'_j})}{\partial x_k} = \frac{\partial(\overline{u'_i}\cdot\overline{u'_j}\cdot\overline{u'_k})}{\partial x_k} - \cancel{(\overline{u'_i}\cdot\overline{u'_j})\frac{\partial\overline{u'_k}}{\partial x_k}}. \quad (5.31)$$

Term T'_2 is rewritten to

$$
\begin{aligned}
& \frac{1}{Re_\tau}\left(\overline{u'_i}\frac{\partial^2\overline{u'_j}}{\partial x_k\partial x_k}+\overline{u'_j}\frac{\partial^2\overline{u'_i}}{\partial x_k\partial x_k}\right) \\
= & \frac{1}{Re_\tau}\left(\frac{\partial}{\partial x_k}\left(\overline{u'_i}\frac{\partial\overline{u'_j}}{\partial x_k}\right)-\frac{\partial\overline{u'_i}}{\partial x_k}\frac{\partial\overline{u'_j}}{\partial x_k}+\frac{\partial}{\partial x_k}\left(\overline{u'_j}\frac{\partial\overline{u'_i}}{\partial x_k}\right)-\frac{\partial\overline{u'_j}}{\partial x_k}\frac{\partial\overline{u'_i}}{\partial x_k}()\right. \\
= & \frac{1}{Re_\tau}\left(-2\frac{\partial\overline{u'_j}}{\partial x_k}\frac{\partial\overline{u'_i}}{\partial x_k}+\frac{\partial^2\overline{u'_i}\cdot\overline{u'_j}}{\partial x_k\partial x_k}\right) \\
= & -\frac{2}{Re_\tau}\left(\frac{\partial\overline{u'_j}}{\partial x_k}\frac{\partial\overline{u'_i}}{\partial x_k}\right)+\frac{1}{Re_\tau}\frac{\partial^2(\overline{u'_i}\cdot\overline{u'_j})}{\partial x_k\partial x_k}.
\end{aligned}
\tag{5.32}
$$

Using

$$
\overline{u'_\ell}\frac{\partial(\overline{u'_k}\langle u_m\rangle)}{\partial x_k}=\overline{u'_\ell}\left(\overline{u'_k}\frac{\partial\langle u_m\rangle}{\partial x_k}+\cancel{\langle u_m\rangle\frac{\partial\overline{u'_k}}{\partial x_k}}\right)=\overline{u'_\ell}\cdot\overline{u'_k}\frac{\partial\langle u_m\rangle}{\partial x_k}
\tag{5.33}
$$

and

$$
\overline{u'_\ell}\frac{\partial(\langle u_k\rangle\overline{u'_m})}{\partial x_k}=\overline{u'_\ell}\left(\langle u_k\rangle\frac{\partial\overline{u'_m}}{\partial x_k}+\cancel{\overline{u'_m}\frac{\partial\langle u_k\rangle}{\partial x_k}}\right)=\overline{u'_\ell}\cdot\langle u_k\rangle\frac{\partial\overline{u'_m}}{\partial x_k}
\tag{5.34}
$$

term T'_3 in equation (5.30) is expressed as

$$
\begin{aligned}
& -\overline{u'_i}\frac{\partial\langle u_k\rangle\overline{u'_j}}{\partial x_k}-\overline{u'_j}\frac{\partial\langle u_k\rangle\overline{u'_i}}{\partial x_k}-\overline{u'_i}\frac{\partial\overline{u'_k}\langle u_j\rangle}{\partial x_k}-\overline{u'_j}\frac{\partial\overline{u'_k}\langle u_i\rangle}{\partial x_k} \\
= & -\overline{u'_i}\cdot\overline{u'_k}\frac{\partial\langle u_j\rangle}{\partial x_k}-\overline{u'_j}\cdot\overline{u'_k}\frac{\partial\langle u_i\rangle}{\partial x_k}-\overline{u'_i}\cdot\langle u_k\rangle\frac{\partial\overline{u'_j}}{\partial x_k}-\overline{u'_j}\cdot\langle u_k\rangle\frac{\partial\overline{u'_i}}{\partial x_k} \\
= & -\overline{u'_i}\cdot\overline{u'_k}\frac{\partial\langle u_j\rangle}{\partial x_k}-\overline{u'_j}\cdot\overline{u'_k}\frac{\partial\langle u_i\rangle}{\partial x_k}-\langle u_k\rangle\left(\overline{u'_i}\frac{\partial\overline{u'_j}}{\partial x_k}+\overline{u'_j}\frac{\partial\overline{u'_i}}{\partial x_k}\right) \\
= & -\overline{u'_i}\cdot\overline{u'_k}\frac{\partial\langle u_j\rangle}{\partial x_k}-\overline{u'_j}\cdot\overline{u'_k}\frac{\partial\langle u_i\rangle}{\partial x_k}-\langle u_k\rangle\frac{\partial(\overline{u'_i}\cdot\overline{u'_j})}{\partial x_k}.
\end{aligned}
\tag{5.35}
$$

Then, expressions (5.31), (5.32), and (5.35) are substituted back into equation (5.30), which results in

$$\begin{aligned}
\frac{\partial(\overline{u_i'}\cdot\overline{u_j'})}{\partial t} &+ \frac{\partial(\overline{u_i'}\cdot\overline{u_j'}\cdot\overline{u_k'})}{\partial x_k} = -\frac{2}{Re_\tau}\left(\frac{\partial\overline{u_j'}}{\partial x_k}\frac{\partial\overline{u_i'}}{\partial x_k}\right) + \frac{1}{Re_\tau}\frac{\partial^2(\overline{u_i'}\cdot\overline{u_j'})}{\partial x_k \partial x_k} \\
&+ \overline{u_i'}\frac{\partial\langle u_k' u_j'\rangle}{\partial x_k} + \overline{u_j'}\frac{\partial\langle u_k' u_i'\rangle}{\partial x_k} \\
&- \overline{u_i'}\cdot\overline{u_k'}\frac{\partial\langle u_j\rangle}{\partial x_k} - \overline{u_j'}\cdot\overline{u_k'}\frac{\partial\langle u_i\rangle}{\partial x_k} - \langle u_k\rangle\frac{\partial(\overline{u_i'}\cdot\overline{u_j'})}{\partial x_k} \\
&- \overline{u_i'}\frac{\partial\tau_{kj}^R}{\partial x_k} - \overline{u_j'}\frac{\partial\tau_{ki}^R}{\partial x_k} - \overline{u_i'}\frac{\partial\overline{p'}}{\partial x_j} - \overline{u_j'}\frac{\partial\overline{p'}}{\partial x_i}. \qquad (5.36)
\end{aligned}$$

Finally, the statistical average of equation (5.36) is the Reynolds stress budget of the filtered velocity field,

$$\begin{aligned}
0 &= \left(\frac{\partial\langle\overline{u_i'}\cdot\overline{u_j'}\rangle}{\partial t} + \langle u_k\rangle\frac{\partial\langle\overline{u_i'}\cdot\overline{u_j'}\rangle}{\partial x_k}\right) - \frac{\partial\langle\overline{u_k'}\cdot\overline{u_i'}\cdot\overline{u_j'}\rangle}{\partial x_k} + \frac{1}{\mathrm{Re}_\tau}\frac{\partial^2\langle\overline{u_i'}\cdot\overline{u_j'}\rangle}{\partial x_k \partial x_k} \\
&- \left\langle\overline{u_i'}\frac{\partial\overline{p'}}{\partial x_j} + \overline{u_j'}\frac{\partial\overline{p'}}{\partial x_i}\right\rangle - \langle\overline{u_i'}\cdot\overline{u_k'}\rangle\frac{\partial\langle u_j\rangle}{\partial x_k} - \langle\overline{u_j'}\cdot\overline{u_k'}\rangle\frac{\partial\langle u_i\rangle}{\partial x_k} \\
&- \frac{2}{\mathrm{Re}_\tau}\left\langle\frac{\partial\overline{u_j'}}{\partial x_k}\frac{\partial\overline{u_i'}}{\partial x_k}\right\rangle - \left\langle\overline{u_i'}\frac{\partial\tau_{kj}^R}{\partial x_k}\right\rangle - \left\langle\overline{u_j'}\frac{\partial\tau_{ki}^R}{\partial x_k}\right\rangle. \qquad (5.37)
\end{aligned}$$

In comparison with the Reynolds stress budget equation, obtained from Pope (2000), page 315, equation (7.178),

$$\begin{aligned}
0 &= -\left(\frac{\partial\langle u_i' u_j'\rangle}{\partial t} + \langle u_k\rangle\frac{\partial\langle u_i' u_j'\rangle}{\partial x_k}\right) - \frac{\partial\langle u_k' u_i' u_j'\rangle}{\partial x_k} + \frac{1}{\mathrm{Re}_\tau}\frac{\partial^2\langle u_i' u_j'\rangle}{\partial x_k \partial x_k} \\
&- \left\langle u_i'\frac{\partial p'}{\partial x_j} + u_j'\frac{\partial p'}{\partial x_i}\right\rangle - \langle u_i' u_k'\rangle\frac{\partial\langle u_j\rangle}{\partial x_k} - \langle u_j' u_k'\rangle\frac{\partial\langle u_i\rangle}{\partial x_k} - \frac{2}{\mathrm{Re}_\tau}\left\langle\frac{\partial u_j'}{\partial x_k}\frac{\partial u_i'}{\partial x_k}\right\rangle, \qquad (5.38)
\end{aligned}$$

additional terms appear, which are caused by the application of the spatial filtering. These terms can be split into a transport term (subfilter-scale diffusion) and a source/sink term for the residual stress (inter-scale energy flux),

$$\left\langle \overline{u'_i}\frac{\partial \tau^R_{kj}}{\partial x_k} + \overline{u'_j}\frac{\partial \tau^R_{ki}}{\partial x_k} \right\rangle = \left\langle \frac{\partial(\overline{u'_i}\tau^R_{kj})}{\partial x_k} - \tau^R_{kj}\frac{\partial \overline{u'_i}}{\partial x_k} + \frac{\partial(\overline{u'_j}\tau^R_{ki})}{\partial x_k} - \tau^R_{ki}\frac{\partial \overline{u'_j}}{\partial x_k} \right\rangle$$
$$= \left(\frac{\partial \langle \overline{u'_i}\tau^R_{kj}\rangle}{\partial x_k} + \frac{\partial \langle \overline{u'_j}\tau^R_{ki}\rangle}{\partial x_k} \right) - \left\langle \tau^R_{kj}\frac{\partial \overline{u'_i}}{\partial x_k} + \tau^R_{ki}\frac{\partial \overline{u'_j}}{\partial x_k} \right\rangle. \quad (5.39)$$

By substituting expression 5.39 back into equation 5.37 and evaluating the equation for the streamwise Reynolds stress component, a transport equation for the streamwise Reynolds stress of the filtered velocity is obtained. For fully-developed turbulent pipe flow this equation reads in cylindrical coordinates

$$\begin{aligned} 0 = & - \frac{1}{r}\frac{\mathrm{d}(r\langle \overline{u'_r}\cdot\overline{u'_z}\cdot\overline{u'_z}\rangle)}{\mathrm{d}r} + \frac{1}{Re_\tau}\frac{1}{r}\frac{\mathrm{d}}{\mathrm{d}r}\left(r\frac{\mathrm{d}\langle \overline{u'_z}\cdot\overline{u'_z}\rangle}{\mathrm{d}r}\right) + 2\left\langle \overline{p'}\frac{\partial \overline{u'_z}}{\partial z}\right\rangle - 2\langle \overline{u'_z}\cdot\overline{u'_r}\rangle\frac{\mathrm{d}\langle u_z\rangle}{\mathrm{d}r} \\ & - \frac{2}{Re_\tau}\left(\left\langle\left(\frac{\partial \overline{u'_z}}{\partial z}\right)^2\right\rangle + \left\langle\frac{1}{r^2}\left(\frac{\partial \overline{u'_z}}{\partial \varphi}\right)^2\right\rangle + \left\langle\left(\frac{\partial \overline{u'_z}}{\partial r}\right)^2\right\rangle\right) \\ & - \frac{2}{r}\frac{\mathrm{d}(r\langle \overline{u'_z}\tau^R_{rz}\rangle)}{\mathrm{d}r} + 2\left(\left\langle \tau^R_{zz}\frac{\partial \overline{u'_z}}{\partial z}\right\rangle + \left\langle \tau^R_{\varphi z}\frac{1}{r}\frac{\partial \overline{u'_z}}{\partial \varphi}\right\rangle + \left\langle \tau^R_{rz}\frac{\partial \overline{u'_z}}{\partial r}\right\rangle\right). \end{aligned} \quad (5.40)$$

The terms in equation 5.40 are denoted as the turbulent transport term (TD), the viscous diffusion (VD) term, the pressure-strain term (PS), the production term (P), the dissipation term (D), the subfilter-scale diffusion term (SD), and the inter-scale energy flux (EF) of the filtered streamwise velocity, when read in order of their appearance. Within the given framework, the terms TD, VD, PS and SD describe the intra-scale energy transport, i.e. the transport of energy of scales larger than the filter length in physical space. The production term in equation (5.40) reflects the amount of streamwise turbulent kinetic energy transferred from the mean field into turbulent coherent structures, which are larger than the filter length. The dissipation term in equation (5.40), on the contrary, depicts the amount of energy directly dissipated from scales larger than the filter length. Finally, the inter-scale energy flux term describes the amount of energy transferred from scales smaller than the filter length into larger ones — or vice versa. Thus, the term acts as a source — or sink — in equation (5.40).

5.3 The Streamwise Reynolds Stress Budget of the Filtered Velocity Field

Recalling the analysis of one-point statistics in section 4.2, the streamwise component contributes most to the turbulent kinetic energy in turbulent pipe flow. As figure 5.3, where turbulent intensities of the unfiltered and the filtered velocity field are displayed, indicates the same is true for the filtered flow field. Particularly, the amount of turbulent kinetic energy carried by VLSMs is dominated by the streamwise component, which justifies that the current study focuses on the transport equation of the streamwise Reynolds stress of the low-pass filtered field (equation 5.40), only. The location of the streamwise turbulent intensity peak in figure 5.3(a) reflects the wall-normal location of the corresponding energy-containing motions. For the unfiltered field, as well as for filter configuration A, the peak is located at $y^+ \approx 15$, where the well-known buffer-layer streaks are the dominant species of coherent structures. For filter configuration B, on the other hand, the peak is located in the upper part of the log layer at $y^+ \approx 390$ ($y/R \approx 0.26$). Figure 5.4 presents the single terms in the budget equation (5.40) for the unfiltered field (a,b), filter configuration A (c,d), and filter configuration B (e,f). The focus is on the near-wall region (figure 5.4a,c,e) as well as on the outer-flow region (figure 5.4b,d,f). Comparing the terms for configuration A (figure 5.4c) with the ones for the unfiltered field (figure 5.4a) in the vicinity of the wall reveals that all TKE budget terms loose in magnitude, when the small scales are removed from the velocity and pressure fields, whereas the terms based on the residual stress appear. The latter are the subfilter-scale diffusion term (yellow line) and the inter-scale energy flux (light-blue line). For filter configuration A, the inter-scale energy flux term is negative at all distances from the wall but the close vicinity to the peak of turbulent production ($y^+ \approx 12$). In this particular region the term exhibits positive values, meaning that more energy is transferred from scales smaller than the filter width to the larger ones than vice versa. This effect, which is referred to as backscattering or inverse energy cascade, has been reported by Härtel et al. (1994) and Piomelli et al. (1996). The latter studies linked the appearance of backscattering to strong shear layers and coherent structures found in the near-wall cycle of wall-bounded turbulence at $Re_\tau = 180$. With increasing Reynolds number, however, additional effect are to be considered. As described in section 4.2.2, outer-flow VLSMs interact with the near-wall cycle, causing a logarithmic Reynolds number dependency of near-wall statistics. In the following, the energy budget related to VLSMs is analysed via filter configuration B. Terms of budget equation (5.40) computed with corresponding filter lengths are presented in figure 5.4(e,f). Unlike for

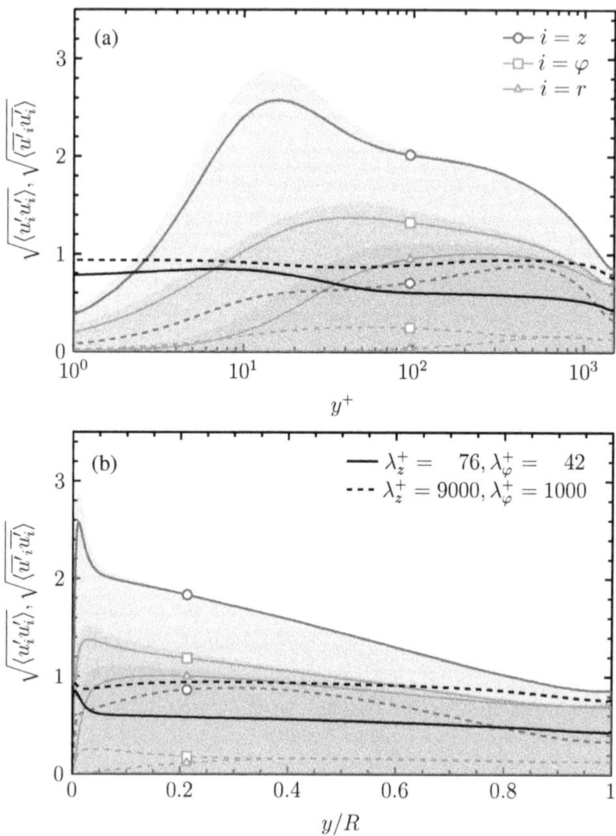

Figure 5.3: Turbulent intensities for the unfiltered field (semi-transparent areas), the low-pass filtered field with $\lambda_z^+ = 76$, $\lambda_\varphi^+ = 42$ (solid lines) and the low-pass filtered field with $\lambda_z^+ = 9000$, $\lambda_\varphi^+ = 1000$ (dashed lines). -○-, $\langle \overline{u'}_z \overline{u'}_z \rangle$; -□-, $\langle \overline{u'}_\varphi \overline{u'}_\varphi \rangle$; -△-, $\langle \overline{u'}_r \overline{u'}_r \rangle$. —, Fraction of the streamwise component with regard to the turbulent kinetic energy $\langle \overline{u'}_z \overline{u'}_z \rangle / \sum_i \langle \overline{u'}_i \overline{u'}_i \rangle$. (a) y-coordinate in wall units; (b) y-coordinate in bulk units.

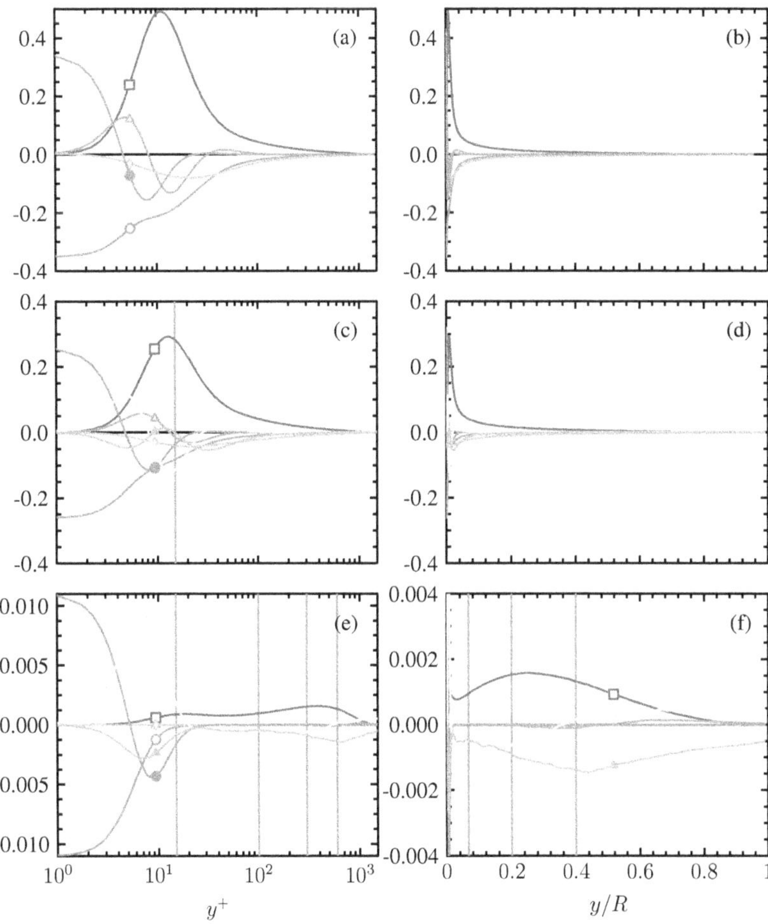

Figure 5.4: Streamwise Reynolds stress budget equation terms for $Re_\tau = 1500$ of the unfiltered field (a,b), the low-pass filtered field with $\lambda_z^+ = 76$, $\lambda_\varphi^+ = 42$ (c,d) and the low-pass filtered field with $\lambda_z^+ = 9000$, $\lambda_\varphi^+ = 1000$ (e,f). -□-, production (P); -○-, dissipation (D); -●-, viscous diffusion (VD); -◇-, pressure-strain (PS); -△-, turbulent diffusion (TD); , subfilter-scale diffusion (SD); -☆-, inter-scale energy flux (EF). All terms normalised in viscous units. Grey lines indicate cutting planes for instantaneous flow field realisations in figures 5.5,5.7,5.8,5.12 and 5.14.

the unfiltered field and filter configuration A, the profiles of the inter-scale energy flux, the subfilter-scale diffusion and the production are of the same order of magnitude throughout most of the flow domain. For the former two cases, on the contrary, all budget terms are most dominant in the vicinity of the wall ($y^+ < 100$). Moreover, viscous diffusion and dissipation only appear near the wall ($y^+ < 20$) and are close to zero in the rest of the flow domain, for filter configuration B. In addition, there is only little contribution to the turbulent diffusion from scales larger than the filter widths of configuration B. While the overall TKE production peaks at $y^+ \approx 12$, the peak for configuration B is located at $y^+ \approx 350$ ($y/R \approx 0.24$). A similar wall distance was reported by Cimarelli et al. (2016) as the location of the outer-scale energy source. For all cases considered — the unfiltered, configuration A, and configuration B —, the production peak is located slightly below the peak of the streamwise turbulent intensity shown in figure 5.3. The large-scale contribution to turbulent production is caused by the large-scale contribution to the Reynolds shear stress, which has also been observed by Ahn et al. (2017). Unlike for filter configuration A, where backscattering appears, the turbulent production term is found to be the only source term in the budget of the VLSMs (figure 5.4f). Particularly the inter-scale energy flux is negative at all distances from the wall. Thus, there is no mean inverse energy cascade for VLSMs. The present study involving filter lengths related to VLSMs expands the findings of previous works by Marati et al. (2004), Cimarelli and De Angelis (2011), and Saikrishnan et al. (2012), which focused on scale energy fluxes up to $\lambda^+ = 250$ using the generalised Kolmogorov equation of the second-order structure function. For these scales backscattering was detected at wall distances of $6 \leq y^+ \leq 37$, which is in agreement with results from filtering configuration A considered here. Albeit a mean inverse cascade is not found for VLSMs, the analysis of instantaneous flow field realisations and conditional statistics carried out in the upcoming section 5.4 reveals that backscattering plays indeed a role for VLSMs. Thus, the current analysis supports the observations of Cimarelli et al. (2013), who found the inverse cascade to be a crucial mechanism in the formation on very long streaks. In physical space the subfilter-scale diffusion term transports energy related to VLSMs away from the peak of its production ($y^+ \approx 350$) towards both the wall and the outer-flow region (yellow line in figure 5.4f). The subfilter-scale diffusion term is further discussed in section 5.5 before the production term is analysed in more detail in section 5.6. The deeper analysis in the upcoming sections involves instantaneous snapshots of the velocity field and the quantities of interest as well as conditional statistics.

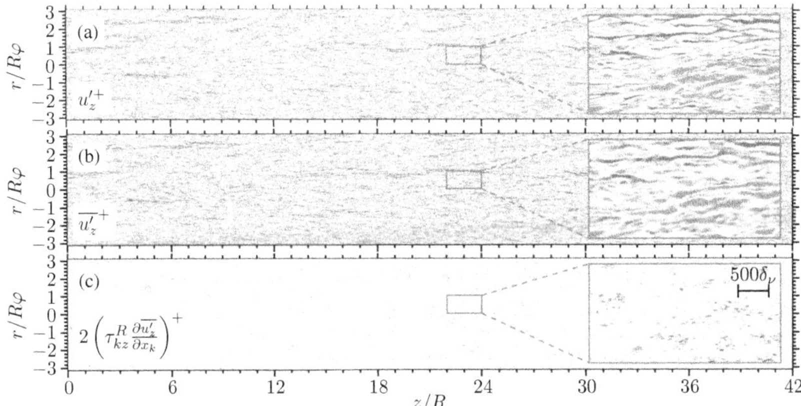

Figure 5.5: Iso-contours in a wall-parallel plane at $y^+ = 15$. (a) Unfiltered streamwise velocity fluctuation u_z'. (b) Low-pass filtered streamwise velocity fluctuation $\overline{u_z'}$. (c) Instantaneous inter-scale energy flux term, $2(\tau_{kz}^R \partial \overline{u_z'}/\partial x_k)^+$. All terms normalised in wall units. Filter lengths: $\lambda_z^+ = 76$, $\lambda_\varphi^+ = 42$. The inset shows a subsection of size $l_z^+ \times l_\varphi^+ = 3000 \times 1500$. (a,b) Values ranging from -5 (dark blue) to 5 (dark red). (c) Values ranging from -2 (dark blue) to 2 (dark red).

5.4 The Inter-scale Energy Flux Term

Using instantaneous flow field realisations and conditional averages, the inter-scale energy flux term of the streamwise budget equation of the low-pass filtered field (5.40) is linked to coherent structures in the velocity field of different size and different sign. Piomelli et al. (1996) already showed the strong correlation between backscattering events and sweeping high-speed fluid in the buffer layer of low Reynolds number wall-bounded turbulence ($Re_\tau = 180$). Forward scattering, on the contrary, was predominately encountered in regions of ejected low-speed fluid. Applying filtering configuration A, which involves filter lengths of the same order of magnitude as used by Piomelli et al. (1996), yields similar results. Unfiltered and configuration A filtered streamwise velocity fluctuations, as well as the instantaneous inter-scale energy flux term for configuration A, are shown in figure 5.5 in a wall-parallel plane at $y^+ = 15$. Comparing figure 5.5(a) with figure 5.5(b) shows, that streamwise velocity fluctuation iso-contours at $y^+ = 15$ do not change significantly when applying filtering configuration A. Particularly, the typical streaky structure of the buffer layer — consisting of small scale wall-layer streaks and VLSMs (see also section 4.1) — is well preserved.

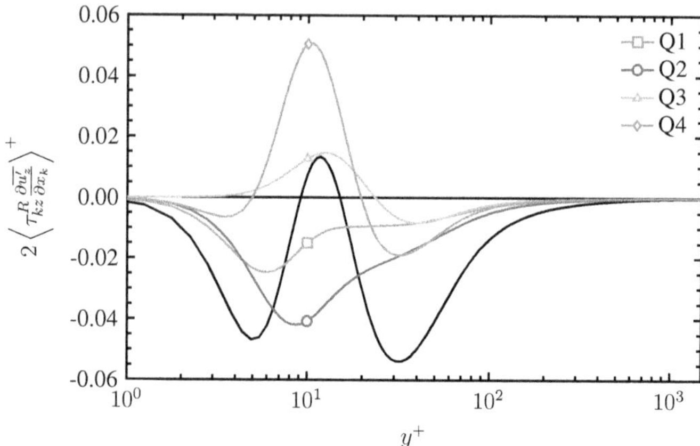

Figure 5.6: Quadrantwise conditionally averaged streamwise inter-scale energy flux term of the low-pass filtered field $2\langle \tau^R_{kz}\partial\overline{u'_z}/\partial x_k\rangle^+$ with filter lengths $\lambda_z^+ = 76$, $\lambda_\varphi^+ = 42$. Conditions: Q1, $\overline{u'_z} > 0$, $\overline{u'_r} < 0$; Q2, $\overline{u'_z} < 0$, $\overline{u'_r} < 0$; Q3, $\overline{u'_z} < 0$, $\overline{u'_r} > 0$; Q4, $\overline{u'_z} > 0$, $\overline{u'_r} > 0$. The black line shows the unconditionally averaged mean inter-scale energy flux as shown in figure 5.4(c).

The structures in the inter-scale energy flux field are much smaller than the above described velocity structures and do not feature the typical streamwise elongation. The correlation discovered by Piomelli et al. (1996) — positive(negative) energy flux with sweeping(ejecting) high(low)-speed fluid — is not easily visible from instantaneous realisations at $Re_\tau = 1500$ (figure 5.5b,c). Hence, the mean inter-scale energy flux profile is computed involving a condition based on the location in the common sample space of the streamwise and the wall-normal velocity fluctuation, depicted in figure 1.9. This procedure allows to associate certain inter-scale energy flux events with certain types of coherent structure statistically. The conditionally averaged inter-scale energy flux for filter configuration A is presented in figure 5.6. While the configuration A inter-scale energy flux profile obtained from Q4 events ($u'_z > 0$, $u'_r > 0$) is positive at $6 < y^+ < 20$ with a peak around $y^+ \approx 10$, the Q2-profile ($u'_z < 0$, $u'_r < 0$) is negative at all wall distances with a minimum around $y^+ \approx 10$ (figure 5.6). Thus, figure 5.6 shows the occurrence of backscattering in regions of high-speed sweeping fluid (or Q4 events) and forward scattering in regions of low-speed ejecting fluid (or Q2 events), which is In good agreement with Piomelli et al. (1996).

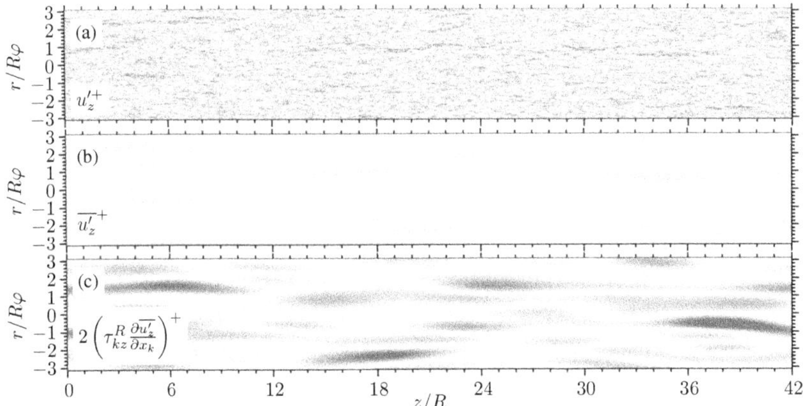

Figure 5.7: Iso-contours in a wall-parallel plane at $y^+ = 15$. (a) Unfiltered streamwise velocity fluctuation u_z'. (b) Low-pass filtered streamwise velocity fluctuation $\overline{u_z'}$. (c) Instantaneous inter-scale energy flux term, $2(\tau^R_{kz}\partial\overline{u_z'}/\partial x_k)^+$. All terms normalised in wall units. Filter lengths: $\lambda_z^+ = 9000$, $\lambda_\varphi^+ = 1000$. (a,b) Values ranging from -5 (dark blue) to 5 (dark red). (c) Values ranging from -0.03 (dark blue) to 0.03 (dark red).

Applying filtering configuration B leads to a drastically different behaviour of the inter-scale energy flux, as will be discussed hereinafter. The fluctuating streamwise velocity field, its configuration B filtered equivalent, and the instantaneous inter-scale energy flux field obtained with configuration B filtering are shown in wall-parallel planes at two distances from the wall in figure 5.7 ($y^+ = 15$) and figure 5.8 ($y^+ = 600$). Close to the wall, SSMs are filtered away and only the VLSM mode remains (figure 5.7a,b). Moreover, figure 5.7(c) displays long regions of both positive and negative instantaneous inter-scale energy flux iso-contours at $y^+ = 15$. Particularly, the appearance of positive inter-scale energy flux events at $y^+ = 15$ for filtering configuration B indicates the occurrence of backscattering towards VLSMs close to the wall. However, positive streamwise velocity fluctuation regions in figure 5.7(b) appear to collapse with negative inter-scale energy flux regions in figure 5.7(c) and vice versa. Thus, unlike for buffer-layer streaks, VLSM backscattering correlates with low-speed fluid, whereas VLSM forward scattering correlates with high-speed structures. In the outer-flow region, the large-scale filtered velocity field (figure 5.8b) strongly resembles the one near the wall (figure 5.7b), whereas the inter-scale energy flux field differs substantially (figure 5.8c and figure 5.7c). At $y^+ = 600$ forward scattering is dominant for both positive and negative VLSMs. The observations regarding the inter-scale energy flux

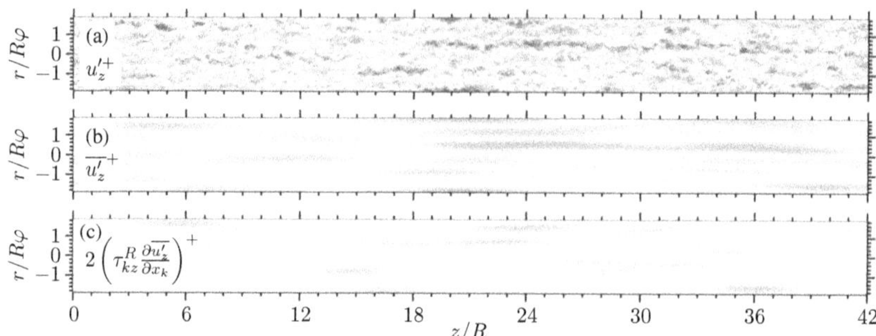

Figure 5.8: Iso-contours in a wall-parallel plane at $y^+ = 600$. (a) Unfiltered streamwise velocity fluctuation u_z'. (b) Low-pass filtered streamwise velocity fluctuation $\overline{u_z'}$. (c) Instantaneous inter-scale energy flux term, $2(\tau_{kz}^R \partial \overline{u_z'} / \partial x_k)^+$. All terms normalised in wall units. Filter lengths: $\lambda_z^+ = 9000$, $\lambda_\varphi^+ = 1000$. (a,b) Values ranging from -5 (dark blue) to 5 (dark red). (c) Values ranging from -0.03 (dark blue) to 0.03 (dark red).

towards and away from VLSMs near the wall as well as in the outer-flow region are again supported by the conditionally averaged inter-scale energy flux profiles depicted in figure 5.9 for filtering configuration B. In contradiction to the effects unveiled via filtering configuration A, the Q2 profile obtained from filtering configuration B (blue line) exhibits backscattering for $y^+ \lesssim 200$, which is most pronounced at a wall distance of $y^+ = 8$, slightly below the configuration A backscattering peak. For filtering configuration B it is the inter-scale energy flux profile related to high-speed sweeping fluid (Q4, red line in figure 5.9), which shows only negative values having a minimum at $y^+ = 8$. Throughout most of the logarithmic layer ($30 \leq y^+ \leq 200$) both profiles remain at a weak constant level, before the Q2-profile turns negative and collapses with the Q4-profile, having a local minimum at $y^+ = 600$. Globally, the forward energy transfer of high-speed sweeping fluid over-compensates the backward transfer of low-speed ejections, which yields no average backscattering into VLSMs at any distance from the wall.

By examination of three-dimensional two-point correlations, i.e.

$$\rho_{ee}(\Delta z, \Delta\varphi, \Delta r, r_0) = \frac{\langle e(z_0, \varphi_0, r_0) e(z_0 + \Delta z, \varphi_0 + \Delta\varphi, r_0 + \Delta r)\rangle}{\langle e^2(z_0, \varphi_0, r_0)\rangle}, \tag{5.41}$$

with $e = \tau_{zk}^R \partial \overline{u_z'} / \partial x_k$, the spatial structure of the inter-scale energy flux can be explored in more detail. Iso-surfaces of this correlation are displayed in figure 5.10

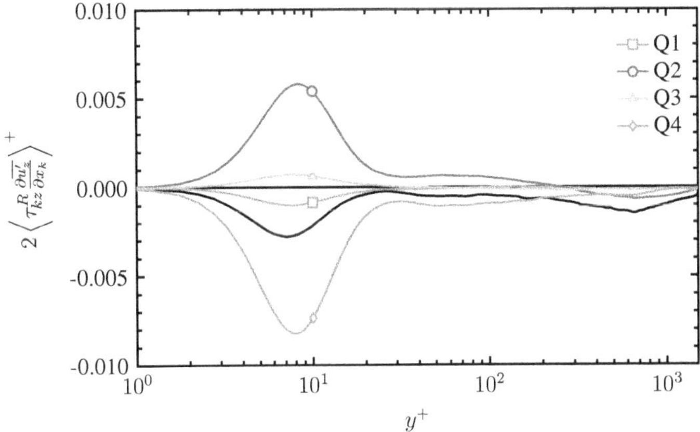

Figure 5.9: Quadrantwise conditionally averaged streamwise inter-scale energy flux term of the low-pass filtered field $2\langle \tau_{kz}^R \partial \overline{u_z'} / \partial x_k \rangle^+$ with filter lengths $\lambda_z^+ = 9000$, $\lambda_\varphi^+ = 1000$. Conditions as in figure 5.6. The black line shows the unconditionally averaged mean inter-scale energy flux as shown in figure 5.4(e).

for filtering configuration A with a reference point $y_0^+ = 15$ and in figure 5.11 for filtering configuration B with a reference point $y/R = 0.4$. The small-scale energy flux structure in figure 5.10 includes an inclined, streamwise elongated structure around the reference point and azimuthally adjacent structures of opposite sign. In terms of the latter features, the small-scale energy flux structure resembles the structure of streamwise velocity fluctuations near the wall (see figure 4.36). However, the inter-scale energy flux structure appears to be smaller in length and width than the velocity structure. Moreover, negative correlated regions above the tail and below the head of the positive energy-flux correlation structure appear in figure 5.10, while being absent in the velocity structure (figure 4.36). The structure of the inter-scale energy flux of VLSMs with reference point at $y_0/R = 0.4$, depicted in figure 5.11, comprises a strong correlated region around the reference point and several strong correlated regions at the wall. The latter are presumably caused by the high intensity of the inter-scale energy flux at the wall. Particularly, the negative correlation "shadow" at the wall below the reference point is due to the sign reversal in the inter-scale energy flux of ejecting VLSMs. Recalling the above discussion regarding figure 5.9, sweeping VLSMs exhibit forward scattering at all distances from the wall, whereas ejecting VLSMs feature backscattering only near the wall. Consequently, the contribution from ejecting

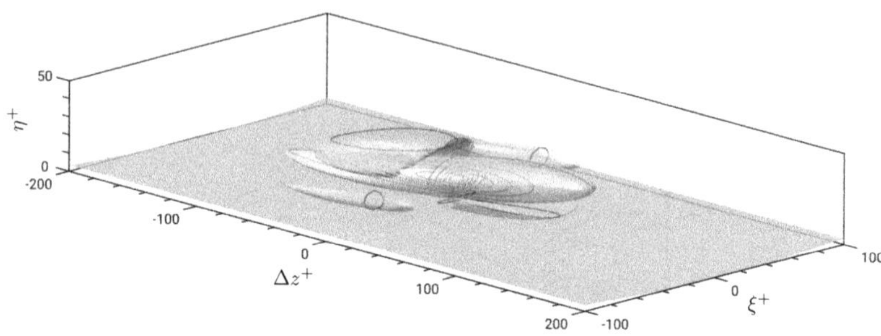

Figure 5.10: Iso-surfaces of the three-dimensional two-point correlation of the streamwise inter-scale energy flux $\rho_{ee}(\Delta z, \Delta\varphi, \Delta r, r_0)$ for configuration A. Reference point in the vicinity of the wall ($(R-r_0)^+ = 15$). Orange (light-blue) iso-surfaces denote a correlation of +(-)0.05. The planes cutting the correlation origin show contour lines ranging from 0.05 to 0.85 (red) and from -0.05 to -0.85 (blue) with an increment of 0.2. Cartesian cross-sectional coordinates $\xi = (r_0 + \Delta r)\sin(\Delta\varphi)$, $\eta = (r_0 + \Delta r)\cos(\Delta\varphi)$.

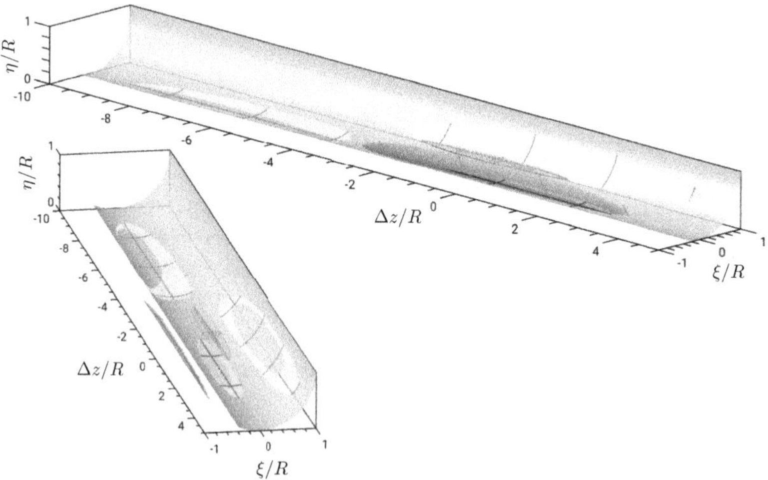

Figure 5.11: Iso-surfaces of the three-dimensional two-point correlation of the streamwise inter-scale energy flux $\rho_{ee}(\Delta z, \Delta\varphi, \Delta r, r_0)$ for configuration B. Reference point in the bulk ($r_0/R = 0.6$). Orange (light-blue) iso-surfaces denote a correlation of +(-)0.7. The planes cutting the correlation origin show contour lines ranging from 0.7 to 0.9 (red) and from -0.7 to -0.9 (blue) with an increment of 0.2. Cartesian cross-sectional coordinates $\xi = (r_0 + \Delta r)\sin(\Delta\varphi)$, $\eta = (r_0 + \Delta r)\cos(\Delta\varphi)$.

VLSMs to the inter-scale energy flux correlation at $y_0/R = 0.4$ reflects in the negative correlation "shadow" described above. Similar to SSMs, VLSMs appear in the shape of azimuthally alternating low-speed ejections and high-speed sweeps (see section 4.3.2), which reflects in the spatial distribution of the correlation "shadows".

5.5 The Subfilter-scale Diffusion Term

The amount of energy related to turbulent scales larger than the filter scale, which is transported by the smaller scales in physical space is represented by the subfilter-scale diffusion term. The instantaneous subfilter-scale diffusion term obtained from both filtering configurations in different wall-parallel planes is presented in figure 5.12. For filtering configuration A, both positive and negative events are found in the region of high-speed fluid in the lower right part of the inset (figure 5.12a, cf. figure 5.5b). The fact that positive and negative events appear in each others close vicinity in the shape of slightly elongated structures can be interpreted as small-scale vortices, which transport the energy of larger scales away from or toward the wall-parallel plane at $y^+ = 15$. Negative events occur more often and are of higher intensity than positive ones, which is consistent with the mean subfilter-scale diffusion profile depicted in figure 5.4(c). Comparing the subfilter-scale diffusion field obtained from filtering configuration B at $y^+ = 15$ (figure 5.12b) with the filtered fluctuating velocity field (figure 5.7b) reveals that positive subfilter-scale diffusion regions coincide with high-speed regions. Thus, energy of VLSM scale is subtracted from low-speed structures and added to high-speed structures by smaller scales. Further away from the wall, the subfilter-scale diffusion field is either relatively uncorrelated (at $y^+ = 100$, figure 5.12c and at $y^+ = 600$, figure 5.12e) or anti-correlated (at $y^+ = 300$, figure 5.12d). In the wall-parallel plane at $y^+ = 300$, which is located in the vicinity of the VLSM production peak ($y^+ \approx 350$), subfilter scales subtract energy from high-speed VLSMs and add it to low-speed VLSMs, opposing the mechanism at the wall distance of $y^+ = 15$. Filtered velocity fields at $y^+ = 100$ and $y^+ = 300$ can be found in appendix F for comparison.

Via conditionally averaged subfilter-scale diffusion profiles, displayed in figure 5.13, the observations from instantaneous fields can be quantified statistically. For filtering configuration A (figure 5.13a), high-speed sweeping fluid contributes most to the mean profile, meaning that the smallest scales mainly distribute energy from the peak of TKE production towards the wall. For filtering configuration B, figure 5.13(b) indicates that sweeps (red curve) and ejections (blue curve) contribute opposingly to the mean profile (black curve). While VLSM sweeps behave similar as SSM sweeps —

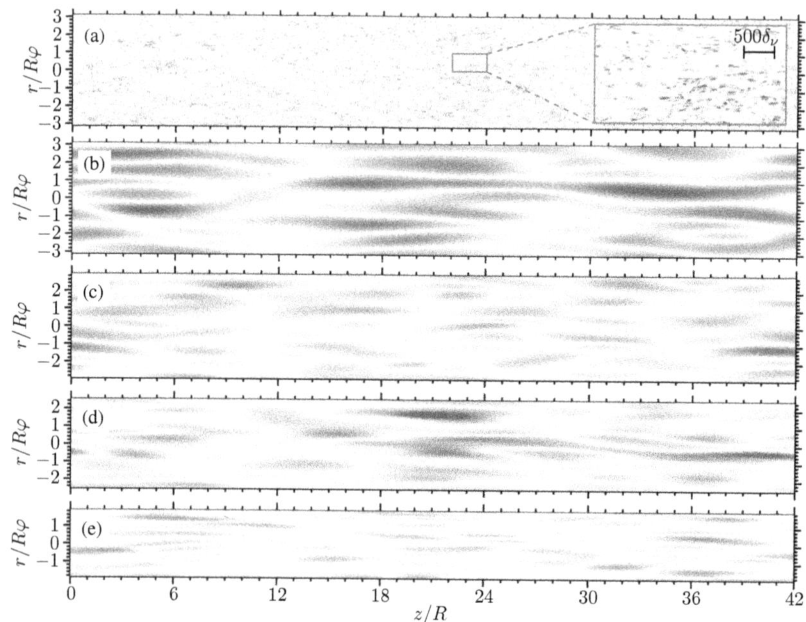

Figure 5.12: Iso-contours of the streamwise instantaneous subfilter-scale diffusion term $-[2/r \cdot \mathrm{d}(r\overline{u'_z \tau^R_{rz}})/\mathrm{d}r]^+$ in wall-parallel planes, normalised in wall units. (a) Filter lengths: $\lambda_z^+ = 76$, $\lambda_\varphi^+ = 42$, $y^+ = 15$.(b,c,d,e) Filter lengths: $\lambda_z^+ = 9000$, $\lambda_\varphi^+ = 1000$. (a,b) $y^+ = 15$. (c) $y^+ = 100$. (d) $y^+ = 300$. (e) $y^+ = 600$. The inset in subfigure (a) shows a subsection of size $l_z^+ \times l_\varphi^+ = 3000 \times 1500$. (a) Values ranging from -2 (dark blue) to 2 (dark red). (b) Values ranging from -0.1 (dark blue) to 0.1 (dark red). (c,d,e) Values ranging from -0.015 (dark blue) to 0.015 (dark red).

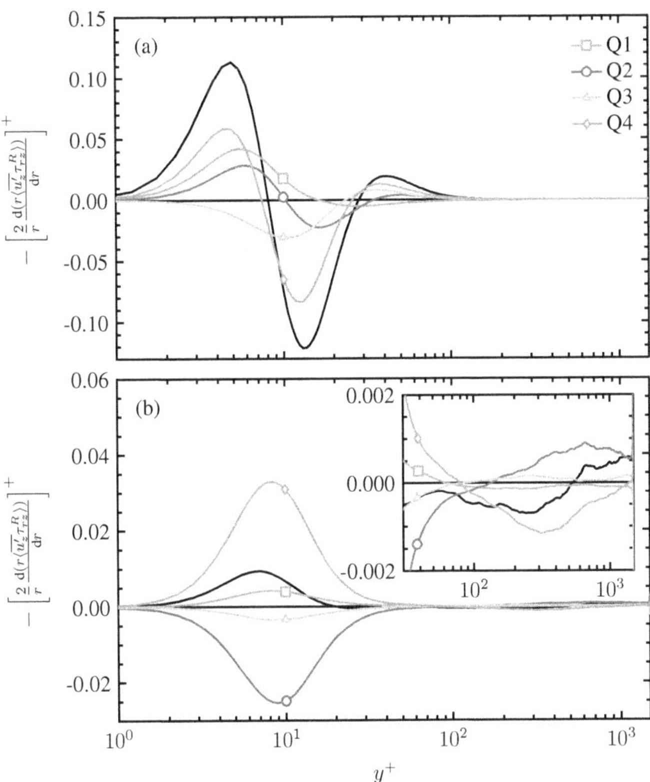

Figure 5.13: Quadrantwise conditionally averaged streamwise subfilter-scale diffusion term $-[2/r \cdot \mathrm{d}(r\langle\overline{u_z' \tau_{rz}^R}\rangle)/\mathrm{d}r]^+$. (a) Filter lengths $\lambda_z^+ = 76$, $\lambda_\varphi^+ = 42$. (b) Filter lengths $\lambda_z^+ = 9000$, $\lambda_\varphi^+ = 1000$. Conditions as in figure 5.6. The black line shows the unconditionally averaged mean subfilter-scale diffusion as shown in figure 5.4(c,e).

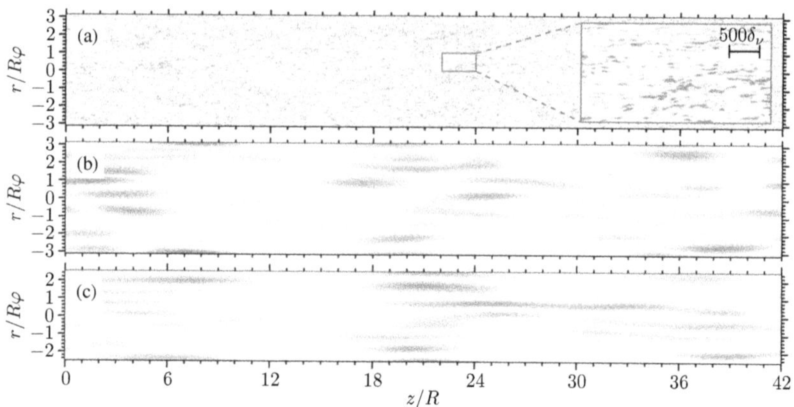

Figure 5.14: Iso-contours of the streamwise instantaneous production term of the filtered field $-2[\overline{u'_z} \cdot \overline{u'_r} \, \mathrm{d}\langle u_z \rangle / \mathrm{d}r]^+$ in wall-parallel planes, normalised in wall units. (a) Filter lengths: $\lambda_z^+ = 76$, $\lambda_\varphi^+ = 42$.(b,c) Filter lengths: $\lambda_z^+ = 9000$, $\lambda_\varphi^+ = 1000$. (a,b) $y^+ = 15$. (c) $y^+ = 300$. The inset in subfigure (a) shows a subsection of size $l_z^+ \times l_\varphi^+ = 3000 \times 1500$. (a) Values ranging from -2 (dark blue) to 2 (dark red). (b,c) Values ranging from -0.015 (dark blue) to 0.015 (dark red).

negative contribution to subfilter-scale diffusion near the TKE production peak and positive contribution near the wall —, VLSM ejections exhibit an opposing behaviour. At $y^+ = 100$, the profiles of high-speed sweeps and low-speed ejections cross each other as they cross zero. In summary, energy is redistributed away from the peak of production towards different wall distances by the subfilter-scale diffusion term for SSMs as well as for VLSMs. The behaviour of the subfilter-scale diffusion for filtering configuration B — adding(subtracting) energy to(from) sweeps(ejections) near the wall, while subtracting(adding) it from(to) them in the bulk — can be explained by large (but still subfilter) scale vortices or vortex packages as the ones introduced by Adrian (2007). Coherently aligning with VLSMs in wall-normal direction, these large rollers lift energy from the wall to the bulk, where low-speed VLSMs are ejected and transport energy from the bulk to the wall, where high-speed VLSMs are swept.

5.6 The Production Term of the Filtered Field

Instantaneous snapshots of the TKE production term of the low-pass filtered field are presented in figure 5.14. For filtering configuration A, the instantaneous instance of the TKE production term is shown at $y^+ = 15$ (figure 5.14a), which is in the vicinity of the corresponding TKE production peak (figure 5.4c). As TKE production is always positive on average, and thus, acts as an energy source in the transport equation of the fluctuating field while being a sink in the transport equation of the mean field, positive events in figure 5.14(a) are much more dominant than negative ones. However, instantaneously negative TKE production events are found in the vicinity of the wall ($y^+ = 15$) for filtering configuration A (figure 5.14a), as well as for filtering configuration B (figure 5.14b), which indicates that the fluctuating field locally energises the mean field as well. For filtering configuration A, the most intensive production events in the lower right corner of the inset in figure 5.14(a) correlate with regions of large streamwise velocity visible in figure 5.5(b). For filtering configuration B, negative spots cover much more of the wall-parallel plane displayed in figure 5.14(b), than for filtering configuration A (figure 5.14a). Hence, VLSMs do not only obtain energy from the mean field via the production mechanism in the vicinity of the wall but they also deliver energy to the mean field. At $y^+ = 300$, in the vicinity of the large-scale TKE production peak, instantaneous TKE production is almost only positive for filtering configuration B, which can be seen in figure 5.14(c). Elongated regions of TKE production events correspond to high- and low-speed VLSMs visible in the configuration B filtered velocity field at the same distance from the wall (see figure F.2 in the appendix). There seems to be no preferable connection between either high- or low-speed VLSMs and an increased rate of TKE production. Again, the above observations are verified statistically via the analysis of conditionally averaged TKE production profiles of the low-pass filtered velocity field, which are shown in figure 5.15. For the small scales (figure 5.15a), high-speed sweeping motions contribute somewhat more to the mean TKE production than low-speed ejections. Inward and outward interactions (Q1 and Q3 events) contribute equally negative to the mean profile. Their contribution is, however, little compared to the contribution of Q2 and Q4 events, yielding an overall positive mean TKE production profile. For filtering configuration B (figure 5.15b), profiles from sweeps and ejections collapse up to $y^+ = 300$. Both profiles exhibit their maximum close to the overall maximum at $y^+ = 350$. In the outer-flow region above $y^+ \approx 300$, TKE production related to ejections (blue curve) contributes marginally more to the overall profile than the TKE production related to sweeps (red curve). Q1 and Q3 events do not play a role for VLSMs.

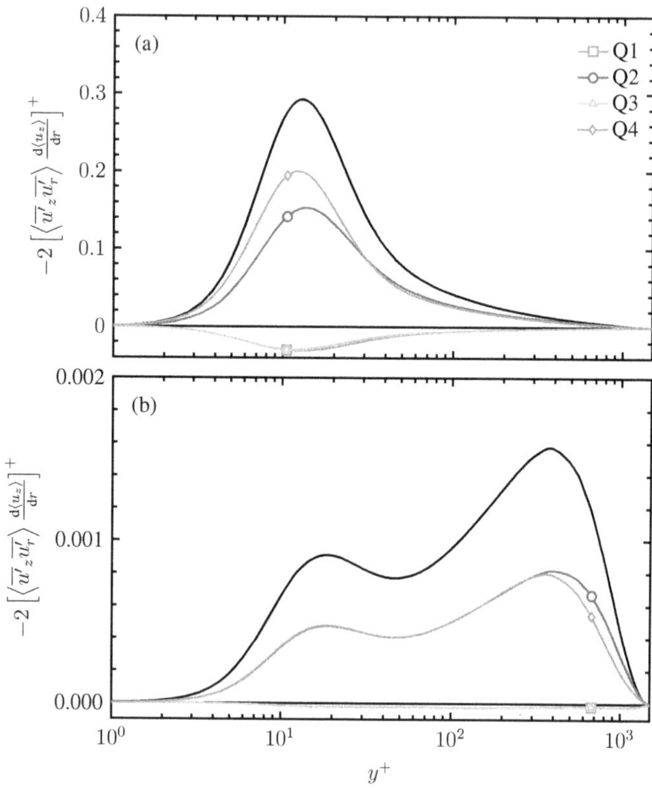

Figure 5.15: Quadrantwise conditionally averaged streamwise production term of the low-pass filtered field $-2[\langle \overline{u'}_z \overline{u'_r} \rangle \, \mathrm{d}\langle u_z \rangle / \mathrm{d}r]^+$. (a) Filter lengths $\lambda_z^+ = 76$, $\lambda_\varphi^+ = 42$. (b) Filter lengths $\lambda_z^+ = 9000$, $\lambda_\varphi^+ = 1000$. Conditions as in figure 5.6. The black line shows the unconditionally averaged mean production of the filtered field as shown in figure 5.4(c,e).

5.7 Concluding Remarks

Energy transfer mechanisms of different scales of coherent motions in turbulent pipe flow were studied via the streamwise Reynolds stress budget equation of the low-pass filtered field. The energy budget of scales comparable to SSMs acknowledges earlier findings from studies of low Reynolds number wall-bounded turbulence (Härtel et al., 1994; Piomelli et al., 1996). As in the latter studies, backscattering of energy into SSMs is detected for a wall distance around $y^+ = 15$. In addition, the traditional terms in the Reynolds stress budget equation obtained with filter lengths $\lambda_z^+ = 76$, $\lambda_\varphi^+ = 42$ act similar to the terms in the unfiltered budget equation. For the budget equation of VLSMs ($\lambda_z^+ = 9000$, $\lambda_\varphi^+ = 1000$), only three dominant terms persist throughout most of the flow domain. These are the production term, the inter-scale energy flux term, and the subfilter-scale diffusion term. Terms involving viscosity — turbulent dissipation and viscous diffusion — only play a role very close to the wall for VLSMs ($y^+ < 20$). With a peak in the upper part of the logarithmic region ($y^+ \approx 350$), the TKE production term is found to be the key source of energy for VLSMs. Subfilter scales are responsible for the transport of VLSM energy away from the production peak location towards the wall and the bulk. Further, the inter-scale energy flux term transfers energy from VLSMs to smaller scales. The first two terms mentioned above act similar as in the case of the small-scale, near-wall cycle with a TKE production peak at $y^+ \approx 12$, where energy is transferred away from by subfilter-scale diffusion beside other transport terms. The inter-scale energy flux term for SSMs is, however, not only dissipative, since energy is transferred from smaller into larger scales at wall distances around $y^+ = 15$.

Earlier observations of Piomelli et al. (1996) that backscattering of energy is related to high-speed sweeping fluid, whereas forward scattering is related to low-speed ejections are confirmed for filter lengths $\lambda_z^+ = 76$, $\lambda_\varphi^+ = 42$ at $Re_\tau = 1500$ via conditional averages. The application of filter lengths $\lambda_z^+ = 9000$, $\lambda_\varphi^+ = 1000$, however, yields opposing results. For VLSMs, the backward energy cascade is correlated with ejections and the forward cascade with sweeps. Since in both cases the energy transfer related to high-speed sweeps is stronger than the one for ejections, on average backscattering is only found for filter lengths $\lambda_z^+ = 76$, $\lambda_\varphi^+ = 42$ in the buffer layer, where SSMs reside. Generally, backscattering events are located in regions of high TKE production. The correlation between the inter-scale energy flux direction and the direction of the fluid motion suggests that those motions are subject to backscattering that enter the region of high TKE production. These are small-scale, high-speed sweeps entering the buffer layer, where TKE production peaks for SSMs on the one hand, and VLSM ejections entering the vicinity of the outer TKE production peak in the logarithmic layer, on the other hand.

As pointed out above, the subfilter-scale diffusion term, representing the amount of large-scale energy transported by smaller scales in physical space, acts similar on both filtering configurations on average. Nonetheless, instantaneous flow field realisations, as well as conditional averaged subfilter-scale diffusion profiles, reveal filter-depended differences. Albeit sweeps contribute somewhat more to subfilter-scale diffusion than ejections for filter lengths $\lambda_z^+ = 76$, $\lambda_\varphi^+ = 42$, both types of coherent structures contribute in a similar way. For filter lengths $\lambda_z^+ = 9000$, $\lambda_\varphi^+ = 1000$, on the contrary, sweeps and ejections contribute to the subfilter-scale diffusion profile in opposite ways. For sweeping VLSMs, subfilter-scale diffusion transports energy away from the VLSM turbulent kinetic energy production peak ($y^+ \approx 350$) towards the wall ($y^+ \approx 8$), whereas for ejecting VLSMs it transports energy from the near-wall region ($y^+ \approx 8$) to the outer-flow region ($y^+ \approx 650$). This behaviour of the subfilter-scale diffusion term in the budget of lengths $\lambda_z^+ = 9000$, $\lambda_\varphi^+ = 1000$ might reflect the interaction between large-scale vortices or vortex packages, which are smaller than the filter scale, and VLSMs being larger than the filter scale.

The above-described behaviour of LSMs and VLSMs resembles the small-scale near-wall cycle, where quasi-streamwise vortices interact with wall-layer streaks in a similar manner. Hwang and Cossu (2010, 2011) and Hwang (2015) showed for turbulent plane channel flow, that large-scale motions self sustain, even after small scales have been artificially removed. Kawata and Alfredsson (2018), by contrast, found an inverse inter-scale transport of Reynolds shear stress in turbulent plane Couette flow, which contributes to TKE production at large scales. The present study affirms the interaction of SSMs and VLSMs in turbulent pipe flow, particularly by means of subfilter-scale diffusion. Moreover, the present work endorses recent findings by Cimarelli et al. (2016) that the overall picture of energy transfer in wall-bounded turbulence involves the coexistence of forward and backward energy transfer at different scales. Via instantaneous observations and conditional averages, particular events of TKE transfer in physical and scale space are connected to particular velocity structures, which expands recent studies on the energy transfer in turbulent plane channel flow by Cho et al. (2018) and Lee and Moser (2019).

Chapter 6

Conclusions

In this research, different aspects of VLSMs in turbulent pipe flow have been analysed with the aid of DNS. The results of this analysis are summarised in the following, referring to the numbering used in section 1.3. As a last point, a future perspective is sketched.

1. The computational domain length L required for the investigation of VLSMs in turbulent pipe flow has been estimated via instantaneous snapshots, turbulence one-point statistics, and spectral statistics. Although low-order statistics are already converged at a much shorter domain length, pre-multiplied energy spectra reveal that capturing a relevant amount of VLSMs demands a domain size of $L \geq 42R$.

2. The convergence of different order moments of the velocity distribution with respect to the near-wall, wall-normal grid spacing Δr_{min} and to the spatio-temporal averaging interval $\Delta\tau$ was studied for $\mathrm{Re}_\tau = 180$. For the grid spacing, a value of $\Delta r^+_{min} = 0.5$ is found to be sufficient for the convergence of the statistical moments of the velocity distribution up to fourth order. The averaging interval required for statistical convergence depends on the order of the statistical moment as well as on the structure of the underlying velocity signal. Averaging intervals of $\Delta\tau^+ = 10^{10}$ for the second-order moments, of $\Delta\tau^+ \approx 2.6 \cdot 10^{11}$ for the third-order moments, and of $\Delta\tau^+ \approx 6.2 \cdot 10^{11}$ for the fourth-order moments are determined. Particularly, the slow convergence of the higher-order moments of the wall-normal velocity distribution in the vicinity of the wall is caused by velocity spikes, the latter being very strong, local, wall-normal velocity fluctuations that appear rarely in space and time.

3. The scaling analysis of statistical moments of the velocity distribution in turbulent pipe flow revealed several logarithmic Reynolds number dependencies. Besides the streamwise Reynolds stress peak, whose logarithmic dependency on Re_τ is well observed for TBL and for plane channel flow, the streamwise velocity skewness and the wall-normal velocity flatness exhibit logarithmic dependencies on the Reynolds number in the vicinity of the wall. Unlike for channel flow, where the logarithmic law is valid from Reynolds numbers as low as $\mathrm{Re}_\tau = 180$, the streamwise Reynolds stress peak in turbulent pipe flow displays a logarithmic behaviour for $\mathrm{Re}_\tau \geq 360$, and the difference between channel and pipe flow is largest at $\mathrm{Re}_\tau = 180$. In the outer-flow region, all flatness components, as well as the spanwise and the wall-normal component of the variance and the skewness, scale with the Reynolds number within the range of the Re_τ considered, whereas the streamwise variance and skewness scale only up to $\mathrm{Re}_\tau = 720$. For $\mathrm{Re}_\tau = 1500$, the streamwise variance increases in the outer-flow region, while the streamwise skewness decreases. The latter behaviour coincides with the appearance of energetic VLSMs at this Reynolds number.

4. Subsequently, the influence of different coherent structures on the above-described scaling behaviour was further elaborated on, after the scaling of coherent structures had been determined via velocity correlations and pre-multiplied energy spectra. Whereas velocity spikes and velocity streaks, as components of the near-wall cycle, scale with the viscous length scale, LSMs and VLSMs scale in bulk units. However, both types of streaky structures of the streamwise velocity fluctuation, buffer-layer streaks and VLSMs, are inclined with regard to the wall. Moreover, for both types of structures, low-speed motions are longer than high-speed motions. In good agreement with Hoyas and Jiménez (2006) and with Mathis et al. (2011), the scaling failure of the streamwise Reynolds stress and of the streamwise velocity skewness in turbulent pipe flow is discovered to be caused by streamwise outer-flow VLSMs interacting with the near wall. Hence, the deletion of these scales from velocity fields leads to Reynolds stress profiles independent of the Reynolds number. Due to the insufficient separation of scales in the Reynolds number regime considered, Reynolds number independence of the streamwise skewness is only obtained very close to the wall ($y^+ \leq 15$) when omitting the correlation term $\langle u'_{zL} u'^2_{zS} \rangle$. As mentioned above, the high wall-normal flatness level at the wall is caused by velocity spikes, which are strongly correlated with high-speed streamwise fluctuations and are, thus, strong sweeps. Consequently, velocity spikes are implicitly redistributed spatially by the interaction of

streamwise VLSMs with near-wall streaks. The interaction is mainly provoked by high-speed streamwise structures, which is determined by conditionally averaged statistics, where the profiles obtained from high-speed structures — i.e. the near-wall profiles of streamwise Reynolds stress and of streamwise skewness as well as the near-wall wall-normal flatness profiles — depend strongly on the Reynolds number. The profiles obtained from low-speed streamwise momentum regions, on the contrary, depend only marginally on the Reynolds number in the vicinity of the wall. In turbulent pipe flow, the wall-normal flatness near the wall at $\mathrm{Re}_\tau = 180$ is significantly larger than in turbulent plane channel flow. This is brought about by the interaction of velocity spikes with the different flow geometries. As velocity spikes scale in viscous units, they "see" the pipe curvature only in the case of low Reynolds numbers. As a result, the discrepancy between channel and pipe flow diminishes for high Reynolds numbers, where the effect of the curvature on viscous flow structures becomes negligibly small.

5. The analysis of the Reynolds stress budget equation of the low-pass filtered field at $\mathrm{Re}_\tau = 1500$ disclosed that similar to SSMs, VLSMs gain their energy via the TKE production mechanism from the mean field. While the SSM production peak is located in the buffer layer at $y^+ \approx 12$, the one related to VLSMs is found in the upper part of the logarithmic layer at $y^+ \approx 350$ ($y/R \approx 0.24$). Unlike SSMs, where a backward transfer of energy is spotted in the close vicinity of the TKE production peak, no mean backscattering of turbulent kinetic energy into VLSMs is detected.

6. While viscous diffusion, turbulent diffusion, pressure strain, subfilter-scale diffusion, production, and turbulent dissipation play a role in the budget of SSMs in addition to the inter-scale energy transfer term, the budget of VLSMs is dominated by the production, the subfilter-scale diffusion, and the inter-scale energy flux term. Only very close to the wall ($y^+ < 20$), where viscous effects are dominant, viscous diffusion plays a role for VLSMs and energy is directly dissipated from VLSMs via turbulent dissipation. The spatial transport of VLSM energy throughout the major part of the flow domain is organised by scales smaller than VLSMs. Therefore, the subfilter-scale diffusion term transports VLSM energy away from the location where it is produced towards both the near-wall and the bulk region. The inter-scale energy flux term acts as a sink in the VLSM budget and is responsible for the transfer of turbulent kinetic energy towards smaller scales.

7. Corresponding with the findings of Piomelli et al. (1996), conditional averages confirmed the correlation between the reverse energy cascade and small-scale high-speed sweeps. The classic forward turbulent cascade, on the other hand, correlates with low-speed ejections. Surprisingly, the effect reverses for VLSMs. For the largest scales of motion in turbulent pipe flow, backscattering correlates with low-speed ejecting fluid and forward scattering with high-speed sweeps. For both SSMs and VLSMs the inter-scale energy transfer peak of sweeps is more intense, so that on average, backscattering is only found for SSMs. In general, backscattering is found in regions of high TKE production. The different correlations between the direction of the inter-scale energy flux and the direction of the fluid motion imply that motions entering these regions of high TKE production are subject to backscattering. For SSMs, these motions are high-speed sweeps entering the buffer layer, whereas for VLSMs, low-speed ejections enter the vicinity of the outer TKE production peak in the logarithmic region.

8. While small-scale sweeps and ejections contribute to subfilter-scale diffusion in a similar manner, the contributions of VLSM sweeps and ejections oppose each other. For sweeping high-speed VLSMs, subfilter-scale diffusion transports energy away from the location of TKE production ($y^+ \approx 350$) towards the wall ($y^+ \approx 8$), whereas for ejecting low-speed VLSMs, it transports energy from the near-wall region ($y^+ \approx 8$) to the outer-flow region ($y^+ \approx 650$). The behaviour of the subfilter-scale diffusion term presumably reflects the effect of large-scale vortices or vortex packages — as the ones described by Adrian (2007) — on VLSMs. These large-scale vortical motions and their alignment with streaky VLSMs resemble the small-scale near-wall cycle, where buffer-layer streaks align coherently with quasi-streamwise vortices, the latter being somewhat smaller in length.

Regarding the dispute whether VLSMs sustain themselves when removing small scales from the simulations of wall-bounded turbulence, as stated by Hwang and Cossu (2010, 2011), the present study rather supports an effect of SSMs on VLSMs. Even though VLSMs are found to be fed with energy by the mean field via TKE production and no backscattering of energy from small scales is detected for VLSMs, the spatial redistribution of turbulent kinetic energy related to VLSMs is carried out by smaller scales.

Overall, the current research work has offered novel insights into the interactions between the viscous small-scale motions in the near-wall region of turbulent pipe flow and outer-flow LSMs and VLSMs. The scaling behaviour of turbulence statistics has been analysed and related to these interactions. Particularly, sources, sinks, and transfer mechanisms of the turbulent kinetic energy of different scales of motions in turbulent pipe flow were identified. However, since VLSMs become more energetic and scales become more separated as the Reynolds number increases, higher Reynolds number DNSs are highly desirable for future investigations of VLSMs in turbulent pipe flow.

As stated above, the current work laid bare the fact that the TKE production mechanism energises VLSMs. Since the production term is a correlation between the Reynolds shear stress and the viscous stress, the question of how the scale-by-scale Reynolds shear stress budget behaves for turbulent pipe flow is of particular interest for future studies. Kawata and Alfredsson (2018) analysed the inter-scale transport of Reynolds shear stress for turbulent plane Couette flow and found an inverse cascade contributing to large-scale structures of Reynolds shear stress. The application of the same approach to the study of turbulent pipe flow is of good potential for future research.

To avoid complications arising from the inhomogeneity in the wall-normal direction, the filtering, as a part of the analysis in chapter 5, is only applied in the two homogeneous directions. Nevertheless, since the turbulence structure consists of a range of scales in all three dimensions, three-dimensional filtering would be worthwhile for the analysis of inter-scale energy flux in the future.

Epilogue

Closing the cycle, the last lines of this thesis shall refer back to its very beginning. Since writing the first lines, many clouds have been advected by the atmospheric boundary layer, the author now finds himself in the 2020 coronavirus lockdown, and the cup of coffee — introduced in chapter 1 — has been replaced by a glass of wine. As I am holding the latter in my hand, another famous quotation of great Professor Feynman comes to my mind. His words shall be the last ones of this thesis. Nothing more to add.

> *A poet once said, "the whole universe is in a glass of wine." We will probably never know in what sense he meant that, for poets do not write to be understood. But it is true that if we look at a glass of wine closely enough we see the entire universe. There are the things of physics: the twisting liquid which evaporates depending on the wind and weather, the reflection in the glass, and our imagination adds the atoms. The glass is a distillation of the earth's rocks, and in its composition we see the secrets of the universe's age, and the evolution of stars. What strange array of chemicals are in the wine? How did they come to be? There are the ferments, the enzymes, the substrates, and the products. There in wine is found the great generalisation: all life is fermentation. Nobody can discover the chemistry of wine without discovering, as did Louis Pasteur, the cause of much disease. How vivid is the claret, pressing its existence into the consciousness that watches it! If our small minds, for some convenience, divide this glass of wine, this universe, into parts — physics, biology, geology, astronomy, psychology, and so on — remember that nature does not know it! So let us put it all back together, not forgetting ultimately what it is for. Let it give us one more final pleasure: drink it and forget it all!* — Feynman et al. (1963)

Bibliography

Abe, H., Kawamura, H., and Choi, H. (2004). Very large-scale structures and their effects on the wall shear-stress fluctuations in a turbulent channel flow up to Re_τ=640. *Journal of Fluids Engineering*, 126(5):835–843.

Abe, H., Kawamura, H., and Matsuo, Y. (2001). Direct numerical simulation of a fully developed turbulent channel flow with respect to the Reynolds number dependence. *Journal of Fluids Engineering*, 123(2):382–393.

Adrian, R. J. (2007). Hairpin vortex organization in wall turbulence. *Physics of Fluids*, 19(4):041301.

Adrian, R. J., Meinhart, C. D., and Tomkins, C. D. (2000). Vortex organization in the outer region of the turbulent boundary layer. *Journal of Fluid Mechanics*, 422:1–54.

Afzal, N. (1982). Fully developed turbulent flow in a pipe: An intermediate layer. *Ingenieur-Archiv*, 52(6):355–377.

Ahn, J., Lee, J., and Sung, H. J. (2017). Contribution of large-scale motions to the Reynolds shear stress in turbulent pipe flows. *International Journal of Heat and Fluid Flow*, 66:209–216.

Ahn, J., Lee, J. H., Jang, S. J., and Sung, H. J. (2013). Direct numerical simulations of fully developed turbulent pipe flows for Re_τ=180, 544 and 934. *International Journal of Heat and Fluid Flow*, 44:222–228.

Ahn, J., Lee, J. H., Lee, J., Kang, J.-h., and Sung, H. J. (2015). Direct numerical simulation of a 30R long turbulent pipe flow at Re_τ=3008. *Physics of Fluids*, 27(6):065110.

Ahn, J. and Sung, H. J. (2017). Relationship between streamwise and azimuthal length scales in a turbulent pipe flow. *Physics of Fluids*, 29(10):105112.

Ansorge, C. and Mellado, J. P. (2014). Global intermittency and collapsing turbulence in the stratified planetary boundary layer. *Boundary-Layer Meteorology*, 153:89–116.

Antonia, R. A., Rajagopalan, S., and Zhu, Y. (1996). Scaling of mean square vorticity in turbulent flows. *Experiments in Fluids*, 20(5):393–394.

Avsarkisov, V., Hoyas, S., Oberlack, M., and García-Galache, J. P. (2014). Turbulent plane Couette flow at moderately high Reynolds number. *Journal of Fluid Mechanics*, 751:R1.

Baidya, R., Baars, W. J., Zimmerman, S., Samie, M., Hearst, R. J., Dogan, E., Mascotelli, L., Zheng, X., Bellani, G., Talamelli, A., Ganapathisubramani, B., Hutchins, N., Marusic, I., Klewicki, J., and Monty, J. P. (2019). Simultaneous skin friction and velocity measurements in high Reynolds number pipe and boundary layer flows. *Journal of Fluid Mechanics*, 871:377–400.

Bailey, S. C. C., Kunkel, G. J., Hultmark, M., Vallikivi, M., Hill, J. P., a. Meyer, K., Tsay, C., Arnold, C. B., and Smits, A. J. (2010). Turbulence measurements using a nanoscale thermal anemometry probe. *Journal of Fluid Mechanics*, 663:160–179.

Bailey, S. C. C. and Smits, A. J. (2010). Experimental investigation of the structure of large- and very-large-scale motions in turbulent pipe flow. *Journal of Fluid Mechanics*, 651:339–356.

Bailon-Cuba, J., Shishkina, O., Wagner, C., and Schumacher, J. (2012). Low-dimensional model of turbulent mixed convection in a complex domain. *Physics of Fluids*, 24(10):107101.

Bardina, J., Ferziger, J., and Rogallo, R. S. (1985). Effect of rotation on isotropic turbulence: Computation and modelling. *Journal of Fluid Mechanics*, 154:321–336.

Barenblatt, G. I. (1993). Scaling laws for fully developed turbulent shear flows. Part 1. Basic hypotheses and analysis. *Journal of Fluid Mechanics*, 248:513–520.

Barenblatt, G. I., Chorin, A. J., and Prostokishin, V. M. (1997). Scaling laws for fully developed turbulent flow in pipes: Discussion of experimental data. *Proceedings of the National Academy of Sciences*, 94(3):773–776.

Barenblatt, G. I. and Prostokishin, V. M. (1993). Scaling laws for fully developed turbulent shear flows. Part 2. Processing of experimental data. *Journal of Fluid Mechanics*, 248:521–529.

Batchelor, G. K. (1967). *An Introduction to Fluid Dynamics.* Cambridge University Press.

Bauer, C., Feldmann, D., and Wagner, C. (2017). On the convergence and scaling of high-order statistical moments in turbulent pipe flow using direct numerical simulations. *Physics of Fluids*, 29(12):125105.

Bauer, C., Feldmann, D., and Wagner, C. (2018). Revisiting the higher-order statistical moments in turbulent pipe flow using direct numerical simulations. In Dillmann, A., Heller, G., Krämer, E., Wagner, C., Bansmer, S., and Radespiel, R., editors, *New Results in Numerical and Experimental Fluid Mechanics XI*, volume 136 of *Notes on Numerical Fluid Mechanics and Multidisciplinary Design*, pages 75–84. Springer International Publishing.

Bauer, C., von Kameke, A., and Wagner, C. (2019). Kinetic energy budget of the largest scales in turbulent pipe flow. *Physical Review Fluids*, 4(6):064607.

Bauer, C. and Wagner, C. (2019a). Analysis of the energy budget of the largest scales in turbulent pipe flow. In Örlü, R., Talamelli, A., Peinke, J., and Oberlack, M., editors, *Progress in Turbulence VIII*, volume 226, pages 113–118. Springer International Publishing.

Bauer, C. and Wagner, C. (2019b). Scaling of high-order statistics in turbulent pipe flow. In Salvetti, M. V., Armenio, V., Fröhlich, J., Geurts, B. J., and Kuerten, H., editors, *Direct and Large-Eddy Simulation XI*, number 25 in ERCOFTAC Series, pages 537–543. Springer International Publishing.

Bernardini, M., Pirozzoli, S., and Orlandi, P. (2014). Velocity statistics in turbulent channel flow up to Re_{τ}=4000. *Journal of Fluid Mechanics*, 742:171–191.

Blackwelder, R. F. and Eckelmann, H. (1979). Streamwise vortices associated with the bursting phenomenon. *Journal of Fluid Mechanics*, 94(3):577–594.

Blasius, H. (1913). Das Ähnlichkeitsgesetz bei Reibungsvorgängen in Flüssigkeiten. *Mitteilungen über Forschungsarbeiten auf dem Gebiete des Ingenieurwesens*, 131:1–41.

Boersma, B. J. (2011). Direct numerical simulation of turbulent pipe flow up to a Reynolds number of 61000. *Journal of Physics: Conference Series*, 318(4):042045.

Boersma, B. J. (2013). Direct numerical simulation of turbulent pipe at high Reynolds numbers, velocity statistics and large scale motions. In *Proc. of the Eighth Symposium on Turbulence and Shear Flow Phenomena, Poitiers, France, August 28-30.*

Boffetta, G. and Ecke, R. E. (2012). Two-dimensional turbulence. *Annual Review of Fluid Mechanics*, 44(1):427–451.

Brodkey, R. S., Wallace, J. M., and Eckelmann, H. (1974). Some properties of truncated turbulence signals in bounded shear flows. *Journal of Fluid Mechanics*, 63(2):209–224.

Brown, G. L. and Thomas, A. S. W. (1977). Large structure in a turbulent boundary layer. *Physics of Fluids*, 20(10):S243.

Calmet, I. and Magnaudet, J. (2003). Statistical structure of high-Reynolds-number turbulence close to the free surface of an open-channel flow. *Journal of Fluid Mechanics*, 474:355–378.

Chin, C., Monty, J. P., and Ooi, A. (2014). Reynolds number effects in DNS of pipe flow and comparison with channels and boundary layers. *International Journal of Heat and Fluid Flow*, 45(1):33–40.

Chin, C., Ooi, A. S. H., Marusic, I., and Blackburn, H. M. (2010). The influence of pipe length on turbulence statistics computed from direct numerical simulation data. *Physics of Fluids*, 22(11):115107.

Cho, M., Hwang, Y., and Choi, H. (2018). Scale interactions and spectral energy transfer in turbulent channel flow. *Journal of Fluid Mechanics*, 854:474–504.

Chorin, A. J. (1967). A numerical method for solving incompressible viscous flow problems. *Journal of Computational Physics*, 2(1):12–26.

Chorin, A. J. (1968). Numerical solution of the Navier-Stokes equations. *Mathematics of Computation*, 22(104):745–762.

Cimarelli, A. and De Angelis, E. (2011). Analysis of the Kolmogorov equation for filtered wall-turbulent flows. *Journal of Fluid Mechanics*, 676:376–395.

Cimarelli, A. and De Angelis, E. (2012). Anisotropic dynamics and sub-grid energy transfer in wall-turbulence. *Physics of Fluids*, 24(1):015102.

Cimarelli, A. and De Angelis, E. (2014). The physics of energy transfer toward improved subgrid-scale models. *Physics of Fluids*, 26(5):055103.

Cimarelli, A., De Angelis, E., and Casciola, C. M. (2013). Paths of energy in turbulent channel flows. *Journal of Fluid Mechanics*, 715:436–451.

Cimarelli, A., De Angelis, E., Jiménez, J., and Casciola, C. M. (2016). Cascades and wall-normal fluxes in turbulent channel flows. *Journal of Fluid Mechanics*, 796:417–436.

Cimarelli, A., De Angelis, E., Schlatter, P., Brethouwer, G., Talamelli, A., and Casciola, C. M. (2015). Sources and fluxes of scale energy in the overlap layer of wall turbulence. *Journal of Fluid Mechanics*, 771:407–423.

Colebrook, C. F. and White, C. M. (1937). Experiments with fluid friction in roughened pipes. *Proceedings of the Royal Society of London. Series A-Mathematical and Physical Sciences*, 161(906):367–381.

Darrigol, O. (2009). *Worlds of Flow: A History of Hydrodynamics from the Bernoullis to Prandtl.* Oxford University Press, Oxford.

del Álamo, J. C. and Jiménez, J. (2003). Spectra of the very large anisotropic scales in turbulent channels. *Physics of Fluids*, 15(6):L41–L44.

del Álamo, J. C., Jiménez, J., Zandonade, P., and Moser, R. D. (2004). Scaling of the energy spectra of turbulent channels. *Journal of Fluid Mechanics*, 500:135–144.

den Toonder, J. M. J. and Nieuwstadt, F. T. M. (1997). Reynolds number effects in a turbulent pipe flow for low to moderate Re. *Physics of Fluids*, 9(11):3398–3409.

Discetti, S., Bellani, G., Örlü, R., Serpieri, J., Vila, C. S., Raiola, M., Zheng, X., Mascotelli, L., Talamelli, A., and Ianiro, A. (2019). Characterization of very-large-scale motions in high-Re pipe flows. *Experimental Thermal and Fluid Science*, 104:1–8.

Durst, F., Jovanovic, J., and Sender, J. (1995a). Detailed measurement of the near-wall region of a turbulent pipe flow. In *Turbulent Shear Flows 9*, pages 225–240. Springer.

Durst, F., Jovanovic, J., and Sender, J. (1995b). LDA measurements in the near-wall region of a turbulent pipe flow. *Journal of Fluid Mechanics*, 295:305.

Eggels, J. G. M., Unger, F., Weiss, M. H., Westerweel, J., Adrian, R. J., Friedrich, R., and Nieuwstadt, F. T. M. (1994). Fully developed turbulent pipe flow: A comparison between direct numerical simulation and experiment. *Journal of Fluid Mechanics*, 268:175–210.

Eggels, J. G. M., Westerweel, J., Nieuwstadt, F. T. M., and Adrian, R. J. (1993). Direct numerical simulation of turbulent pipe flow. *Applied Scientific Research*, 51(1-2):319–324.

El Khoury, G. K., Schlatter, P., Noorani, A., Fischer, P. F., Brethouwer, G., and Johansson, A. V. (2013). Direct numerical simulation of turbulent pipe flow at moderately high Reynolds numbers. *Flow, Turbulence and Combustion*, 91(3):475–495.

Feldmann, D. (2015). *Eine numerische Studie zur turbulenten Bewegungsform in der oszillierenden Rohrströmung.* Dissertation, Technische Universität Ilmenau.

Feldmann, D., Bauer, C., and Wagner, C. (2018). Computational domain length and Reynolds number effects on large-scale coherent motions in turbulent pipe flow. *Journal of Turbulence*, 19(3):274–295.

Feldmann, D. and Wagner, C. (2012). Direct numerical simulation of fully developed turbulent and oscillatory pipe flows at Re_τ=1440. *Journal of Turbulence*, 13(32):1–28.

Feynman, R. P., Leighton, R. B., and Sands, M. (1963). *The Feynman Lectures on Physics*, volume 1. Addison-Wesley Publishing Company.

Fiorini, T. (2017). *Turbulent Pipe Flow-High Resolution Measurements in CICLoPE.* Dissertation, Universita di Bologna.

Frisch, U. (1995). *Turbulence: The Legacy of A.N. Kolmogorov.* Cambridge University Press.

Furuichi, N., Terao, Y., Wada, Y., and Tsuji, Y. (2015). Friction factor and mean velocity profile for pipe flow at high Reynolds numbers. *Physics of Fluids*, 27(9):095108.

Gavrilakis, S. (1992). Numerical simulation of low-Reynolds-number turbulent flow through a straight square duct. *Journal of Fluid Mechanics*, 244:101–129.

Grötzbach, G. and Schumann, U. (1979). Direct numerical simulation of turbulent velocity, pressure, and temperature fields in channel flows. In Durst, F., Launder, B. E., Schmidt, F. W., and Whitelaw, J. H., editors, *Turbulent Shear Flows I*, pages 370–385.

Guala, M., Hommema, S. E., and Adrian, R. J. (2006). Large-scale and very-large-scale motions in turbulent pipe flow. *Journal of Fluid Mechanics*, 554:521–542.

Handler, R. A., Saylor, J. R., Leighton, R. I., and Rovelstad, A. L. (1999). Transport of a passive scalar at a shear-free boundary in fully developed turbulent open channel flow. *Physics of Fluids (1994-present)*, 11(9):2607–2625.

Handler, R. A., Swean, T. F., Leighton, R. I., and Swearingen, J. D. (1993). Length scales and the energy balance for turbulence near a free surface. *AIAA Journal*, 31(11):1998–2007.

Härtel, C., Kleiser, L., Unger, F., and Friedrich, R. (1994). Subgrid-scale energy transfer in the near-wall region of turbulent flows. *Physics of Fluids*, 6(9):3130–3143.

Head, M. R. and Bandyopadhyay, P. (1981). New aspects of turbulent boundary-layer structure. *Journal of Fluid Mechanics*, 107:297.

Hellström, L. H. O., Sinha, A., and Smits, A. J. (2011). Visualizing the very-large-scale motions in turbulent pipe flow. *Physics of Fluids*, 23(1):011703.

Hellström, L. H. O. and Smits, A. J. (2014). The energetic motions in turbulent pipe flow. *Physics of Fluids*, 26(12):125102.

Hill, R. J. (2002). Exact second-order structure-function relationships. *Journal of Fluid Mechanics*, 468:317–326.

Horn, S., Shishkina, O., and Wagner, C. (2013). On non-Oberbeck–Boussinesq effects in three-dimensional Rayleigh–Bénard convection in glycerol. *Journal of Fluid Mechanics*, 724:175–202.

Hoyas, S. and Jiménez, J. (2006). Scaling of the velocity fluctuations in turbulent channels up to Re_{τ}=2003. *Physics of Fluids*, 18(1):11702.

Hoyas, S. and Jiménez, J. (2008). Reynolds number effects on the Reynolds-stress budgets in turbulent channels. *Physics of Fluids*, 20(10):101511.

Hultmark, M., Vallikivi, M., Bailey, S. C. C., and Smits, A. J. (2012). Turbulent pipe flow at extreme Reynolds numbers. *Physical Review Letters*, 108(9):094501.

Hutchins, N. and Marusic, I. (2007a). Evidence of very long meandering features in the logarithmic region of turbulent boundary layers. *Journal of Fluid Mechanics*, 579:1–28.

Hutchins, N. and Marusic, I. (2007b). Large-scale influences in near-wall turbulence. *Philosophical Transactions of the Royal Society A: Mathematical, Physical and Engineering Sciences*, 365(1852):647–664.

Hwang, Y. (2015). Statistical structure of self-sustaining attached eddies in turbulent channel flow. *Journal of Fluid Mechanics*, 767:254–289.

Hwang, Y. and Cossu, C. (2010). Self-sustained process at large scales in turbulent channel flow. *Physical Review Letters*, 105(4):044505.

Hwang, Y. and Cossu, C. (2011). Self-sustained processes in the logarithmic layer of turbulent channel flows. *Physics of Fluids*, 23(6):061702.

Jeong, J., Hussain, F., Schoppa, W., and Kim, J. (1997). Coherent structures near the wall in a turbulent channel flow. *Journal of Fluid Mechanics*, 332:185–214.

Jiménez, J. (1998). The largest scales of turbulent wall flows. Technical report, CTR Annual Research Briefs.

Jiménez, J. (2012). Cascades in wall-bounded turbulence. *Annual Review of Fluid Mechanics*, 44(1):27–45.

Jiménez, J., del Álamo, J. C., and Flores, O. (2004). The large-scale dynamics of near-wall turbulence. *Journal of Fluid Mechanics*, 505:179–199.

Jiménez, J. and Hoyas, S. (2008). Turbulent fluctuations above the buffer layer of wall-bounded flows. *Journal of Fluid Mechanics*, 611:215–236.

Jiménez, J., Kawahara, G., Simens, M. P., Nagata, M., and Shiba, M. (2005). Characterization of near-wall turbulence in terms of equilibrium and "bursting" solutions. *Physics of Fluids*, 17(1):15105.

Jiménez, J. and Moin, P. (1991). The minimal flow unit in near-wall turbulence. *Journal of Fluid Mechanics*, 225:213–240.

Jiménez, J. and Pinelli, A. (1999). The autonomous cycle of near-wall turbulence. *Journal of Fluid Mechanics*, 389:335–359.

Kaczorowski, M. and Wagner, C. (2009). Analysis of the thermal plumes in turbulent Rayleigh–Bénard convection based on well-resolved numerical simulations. *Journal of Fluid Mechanics*, 618:89–112.

Kawata, T. and Alfredsson, P. H. (2018). Inverse interscale transport of the Reynolds shear stress in plane Couette turbulence. *Physical Review Letters*, 120(24):244501.

Kim, J. and Moin, P. (1985). Application of a fractional-step method to incompressible Navier-Stokes equations. *Journal of Computational Physics*, 59(2):308–323.

Kim, J., Moin, P., and Moser, R. (1987). Turbulence statistics in fully developed channel flow at low Reynolds number. *Journal of Fluid Mechanics*, 177:133–166.

Kim, K. C. and Adrian, R. J. (1999). Very large-scale motion in the outer layer. *Physics of Fluids*, 11(2):417–422.

Kline, S. J., Reynolds, W. C., a. Schraub, F., and Runstadler, P. W. (1967). The structure of turbulent boundary layers. *Journal of Fluid Mechanics*, 30(4):741.

Kolmogorov, A. N. (1933). *Grundbegriffe der Wahrscheinlichkeitsrechnung*. Springer.

Kolmogorov, A. N. (1941). The local structure of turbulence in incompressible viscous fluid for very large Reynolds numbers. *Doklady Akademii Nauk SSSR*, 30(1890):301–305.

König, F., Zanoun, E.-S., Öngüner, E., and Egbers, C. (2014). The CoLaPipe - The new Cottbus large pipe test facility at Brandenburg University of Technology Cottbus-Senftenberg. *Review of Scientific Instruments*, 85(7):075115.

Kovasznay, L. S. G., Kibens, V., and Blackwelder, R. F. (1970). Large-scale motion in the intermittent region of a turbulent boundary layer. *Journal of Fluid Mechanics*, 41(2):283–325.

Kraichnan, R. H. (1967). Inertial ranges in two-dimensional turbulence. *Physics of Fluids*, 10(7):1417.

Kundu, P. K. and Cohen, I. M. (2008). *Fluid Mechanics*. Academic Press.

Laufer, J. (1954). The structure of turbulence in fully developed pipe flow. NACA Report 1174, NACA.

Lawn, C. J. (1971). The determination of the rate of dissipation in turbulent pipe flow. *Journal of Fluid Mechanics*, 48(3):477–505.

Lee, J., Ahn, J., and Sung, H. J. (2015). Comparison of large- and very-large-scale motions in turbulent pipe and channel flows. *Physics of Fluids*, 27(2):025101.

Lee, J. H. and Sung, H. J. (2013). Comparison of very-large-scale motions of turbulent pipe and boundary layer simulations. *Physics of Fluids*, 25(4):045103.

Lee, M. and Moser, R. D. (2015). Direct numerical simulation of turbulent channel flow up to Re_τ=5200. *Journal of Fluid Mechanics*, 774:395–415.

Lee, M. and Moser, R. D. (2018). Extreme-scale motions in turbulent plane Couette flows. *Journal of Fluid Mechanics*, 842:128–145.

Lee, M. and Moser, R. D. (2019). Spectral analysis of the budget equation in turbulent channel flows at high Reynolds number. *Journal of Fluid Mechanics*, 860:886–938.

Lee, M. J., Kim, J., and Moin, P. (1990). Structure of turbulence at high shear rate. *Journal of Fluid Mechanics*, 216:561–583.

Lenaers, P., Li, Q., Brethouwer, G., Schlatter, P., and Örlü, R. (2012). Rare backflow and extreme wall-normal velocity fluctuations in near-wall turbulence. *Physics of Fluids*, 24(3):035110.

Luchini, P. (2017). Universality of the Turbulent Velocity Profile. *Physical Review Letters*, 118(22):224501.

Marati, N., Casciola, C. M., and Piva, R. (2004). Energy cascade and spatial fluxes in wall turbulence. *Journal of Fluid Mechanics*, 521:191–215.

Marusic, I., Monty, J. P., Hultmark, M., and Smits, A. J. (2013). On the logarithmic region in wall turbulence. *Journal of Fluid Mechanics*, 716:R3.

Mathieu, J. and Scott, J. (2000). *An Introduction to Turbulent Flow*. Cambridge University Press.

Mathis, R., Hutchins, N., and Marusic, I. (2009a). Large-scale amplitude modulation of the small-scale structures in turbulent boundary layers. *Journal of Fluid Mechanics*, 628:311–337.

Mathis, R., Marusic, I., Hutchins, N., and Sreenivasan, K. R. (2011). The relationship between the velocity skewness and the amplitude modulation of the small scale by the large scale in turbulent boundary layers. *Physics of Fluids*, 23(12):121702.

Mathis, R., Monty, J. P., Hutchins, N., and Marusic, I. (2009b). Comparison of large-scale amplitude modulation in turbulent boundary layers, pipes, and channel flows. *Physics of Fluids*, 21(11):111703.

McKeon, B. J., Li, J., Jiang, W., Morrison, J. F., and Smits, A. J. (2004a). Further observations on the mean velocity distribution in fully developed pipe flow. *Journal of Fluid Mechanics*, 501:135–147.

McKeon, B. J., Swanson, C. J., Zagarola, M. V., Donnelly, R. J., and Smits, A. J. (2004b). Friction factors for smooth pipe flow. *Journal of Fluid Mechanics*, 511:41–44.

McKeon, B. J., Zagarola, M. V., and Smits, A. J. (2005). A new friction factor relationship for fully developed pipe flow. *Journal of Fluid Mechanics*, 538:429–443.

Millikan, C. B. (1938). A critical discussion of turbulent flows in channels and circular tubes. In *Proc. 5th Int. Congr. Appl. Mech*, pages 386–392. Wiley.

Moin, P. and Kim, J. (1982). Numerical investigation of turbulent channel flow. *Journal of Fluid Mechanics*, 118:341–377.

Monin, A. S. and Yaglom, A. (1971). *Statistical Fluid Mechanics: Mechanics of Turbulence*, volume 1. MIT Press.

Monin, A. S. and Yaglom, A. (1975). *Statistical Fluid Mechanics: Mechanics of Turbulence*, volume 2. MIT Press.

Monty, J. P., Hutchins, N., Ng, H. C. H., Marusic, I., and Chong, M. S. (2009). A comparison of turbulent pipe, channel and boundary layer flows. *Journal of Fluid Mechanics*, 632:431–442.

Monty, J. P., Stewart, J. A., Williams, R. C., and Chong, M. S. (2007). Large-scale features in turbulent pipe and channel flows. *Journal of Fluid Mechanics*, 589:147–156.

Moody, L. F. (1944). Friction factors for pipe flow. *Trans. ASME*, 66:671–684.

Morrison, J. F., McKeon, B. J., Jiang, W., and Smits, A. J. (2004). Scaling of the streamwise velocity component in turbulent pipe flow. *Journal of Fluid Mechanics*, 508:99–131.

Moser, R. D., Kim, J., and Mansour, N. N. (1999). Direct numerical simulation of turbulent channel flow up to Re_τ=590. *Physics of Fluids*, 11(4):943–945.

Murlis, J., Tsai, H. M., and Bradshaw, P. (1982). The structure of turbulent boundary layers at low Reynolds numbers. *Journal of Fluid Mechanics*, 122:13–56.

Nagaosa, R. (1999). Direct numerical simulation of vortex structures and turbulent scalar transfer across a free surface in a fully developed turbulence. *Physics of Fluids*, 11(6):1581–1595.

Nagaosa, R. and Handler, R. A. (2003). Statistical analysis of coherent vortices near a free surface in a fully developed turbulence. *Physics of Fluids*, 15(2):375–394.

Nagib, H. M. and Chauhan, K. A. (2008). Variations of von Kármán coefficient in canonical flows. *Physics of Fluids*, 20(10):101518.

Navier, C. L. M. H. (1823). Mémoire sur les lois du mouvement des fluides. *Mem. Acad. Sci. Inst. Fr*, 6:389–416.

Nikuradse, J. (1932). Gesetzmässigkeiten der turbulenten Strömung in glatten Rohren. *Forschungsheft [auf d. Geb. d. Ingenieurwesens]*, 356.

Nikuradse, J. (1933). Strömungsgesetze in rauhen Rohren. *Forschungsheft [auf d. Geb. d. Ingenieurwesens]*, 361.

O'Neill, P. L., Nicolaides, D., Honnery, D., Soria, J., and Others (2004). Autocorrelation functions and the determination of integral length with reference to experimental and numerical data. In *15th Australasian Fluid Mechanics Conference*, volume 1, pages 1–4. The University of Sydney.

Öngüner, E. (2018). *Experiments in Pipe Flows at Transitional and Very High Reynolds Numbers*. Dissertation, Brandenburgische Technische Universität, Cottbus-Senftenberg.

Örlü, R., Fiorini, T., Segalini, A., Bellani, G., Talamelli, A., and Alfredsson, P. H. (2017). Reynolds stress scaling in pipe flow turbulence—first results from CICLoPE. *Philosophical Transactions of the Royal Society A: Mathematical, Physical and Engineering Sciences*, 375(2089):20160187.

Orszag, S. A. (1970). Analytical theories of turbulence. *Journal of Fluid Mechanics*, 41(2):363–386.

Orszag, S. A. and Patterson, G. S. (1972). Numerical simulation of three-dimensional homogeneous isotropic turbulence. *Physical Review Letters*, 28(2):76–79.

Panton, R. L. (2007). Composite asymptotic expansions and scaling wall turbulence. *Philosophical Transactions of the Royal Society A: Mathematical, Physical and Engineering Sciences*, 365(1852):733–754.

Panton, R. L. (2009). Scaling and correlation of vorticity fluctuations in turbulent channels. *Physics of Fluids*, 21(11):115104.

Patel, V. C. and Head, M. (1969). Some observations on skin friction and velocity profiles in fully developed pipe and channel flows. *Journal of Fluid Mechanics*, 38(1):181–201.

Perry, A. E. and Abell, C. J. (1975). Scaling laws for pipe-flow turbulence. *Journal of Fluid Mechanics*, 67(2):257–271.

Perry, A. E. and Chong, M. S. (1982). On the mechanism of wall turbulence. *Journal of Fluid Mechanics*, 119:173–217.

Perry, a. E., Henbest, S., and Chong, M. S. (1986). A theoretical and experimental study of wall turbulence. *Journal of Fluid Mechanics*, 165:163–199.

Perry, A. E. and Marušic, I. (1995). A wall-wake model for the turbulence structure of boundary layers. Part 1. Extension of the attached eddy hypothesis. *Journal of Fluid Mechanics*, 298:361.

Pestana, T. and Hickel, S. (2019). Regime transition in the energy cascade of rotating turbulence. *Physical Review E*, 99(5):053103.

Piomelli, U., Yu, Y., and Adrian, R. J. (1996). Subgrid-scale energy transfer and near-wall turbulence structure. *Physics of Fluids*, 8(1):215–224.

Pirozzoli, S., Bernardini, M., and Orlandi, P. (2014). Turbulence statistics in Couette flow at high Reynolds number. *Journal of Fluid Mechanics*, 758:327–343.

Pope, S. B. (2000). *Turbulent Flows*. Cambridge University Press.

Prandtl, L. (1925). Bericht über Untersuchungen zur ausgebildeten Turbulenz. *ZAMM - Journal of Applied Mathematics and Mechanics / Zeitschrift für Angewandte Mathematik und Mechanik*, 5(2):136–139.

Reynolds, O. (1883). An experimental investigation of the circumstances which determine whether the motion of water shall be direct or sinuous, and of the law of resistance in parallel channels. *Proceedings of the royal society of London*, 35(224-226):84–99.

Reynolds, O. (1894). On the dynamical theory of incompressible viscous fluids and the determination of the criterion. *Proceedings of the Royal Society of London*, 56(336-339):40–45.

Richardson, L. F. (1922). *Weather Prediction by Numerical Process*. Cambridge University Press.

Robinson, S. K. (1991). Coherent motions in the turbulent boundary layer. *Annual Review of Fluid Mechanics*, 23(1):601–639.

Rogers, M. M. and Moin, P. (1987). The structure of the vorticity field in homogeneous turbulent flows. *Journal of Fluid Mechanics*, 176:33–66.

Rosenberg, B. J., Hultmark, M., Vallikivi, M., Bailey, S. C. C., and Smits, A. J. (2013). Turbulence spectra in smooth- and rough-wall pipe flow at extreme Reynolds numbers. *Journal of Fluid Mechanics*, 731:46–63.

Rouse, H. (1943). Evaluation of boundary roughness. In *Proc. of the II Hydraulics Conference of Iowa*, volume 27, pages 105–116. University of Iowa studies in engineering, Iowa City.

Sabot, J. and Comte-Bellot, G. (1976). Intermittency of coherent structures in the core region of fully developed turbulent pipe flow. *Journal of Fluid Mechanics*, 74(4):767–796.

Saikrishnan, N., De Angelis, E., Longmire, E., Marusic, I., Casciola, C., and Piva, R. (2012). Reynolds number effects on scale energy balance in wall turbulence. *Physics of Fluids*, 24(1):015101.

Sakai, Y. (2019). *Coherent Structures and Secondary Motions in Open Duct Flow*. Dissertation, Karlsruher Institut für Technologie.

Schlichting, H. and Gersten, K. (2017). *Boundary-Layer Theory*. Springer, ninth edition.

Schmitt, L. and Friedrich, R. (1982). Numerische Simulation turbulenter Grenzschichten (Large-Eddy-Simulation). Technical report, Institut für Strömungsmechanik, Munich.

Schmitt, L. and Friedrich, R. (1988). Large eddy simulation of turbulent backward facing step flow. In *Notes on Numerical Fluid Mechanics*, volume 20, pages 355–362.

Schmitt, L., Richter, K., and Friedrich, R. (1986). Large-eddy simulation of turbulent boundary layer and channel flow at high Reynolds number. In Schumann, U. and Friedrich, R., editors, *Direct and Large Eddy Simulation of Turbulence*, volume 3 of *Notes on Numerical Fluid Mechanics*, pages 161–176. Vieweg+Teubner Verlag.

Schumann, U. (1973). *Ein Verfahren zur direkten numerischen Simulation turbulenter Strömungen in Platten und Ringspaltkanälen und über seine Anwendung zur Untersuchung von Turbulenzmodellen.* Dissertation, Universität Karlsruhe (TH).

Schumann, U. (1975). Subgrid scale model for finite difference simulations of turbulent flows in plane channels and annuli. *Journal of Computational Physics*, 18(4):376–404.

Schumann, U., Grötzbach, G., and Kleiser, L. (1979). Direct numerical simulation of turbulence. In *Prediction Methods for Turbulent Flows*, VKI-Lecture Series, Brussels.

Shishkina, O., Shishkin, A., and Wagner, C. (2009). Simulation of turbulent thermal convection in complicated domains. *Journal of Computational and Applied Mathematics*, 226(2):336–344.

Shishkina, O. and Wagner, C. (2004). Stability conditions for the Leapfrog-Euler scheme with central spatial discretization of any order. *Applied Numerical Analysis and Computational Mathematics*, 1(1):315–326.

Shishkina, O. and Wagner, C. (2005). A fourth order accurate finite volume scheme for numerical simulations of turbulent Rayleigh–Bénard convection in cylindrical containers. *Comptes Rendus Mécanique*, 333:17–28.

Shishkina, O. and Wagner, C. (2007). A fourth order finite volume scheme for turbulent flow simulations in cylindrical domains. *Computers and Fluids*, 36(2):484–497.

Sillero, J. A., Jiménez, J., and Moser, R. D. (2013). One-point statistics for turbulent wall-bounded flows at Reynolds numbers up to $delta^+ approx 2000$. *Physics of Fluids*, 25(10):105102.

Sillero, J. A., Jiménez, J., and Moser, R. D. (2014). Two-point statistics for turbulent boundary layers and channels at Reynolds numbers up to $delta^+ approx 2000$. *Physics of Fluids*, 26(10):105109.

Smits, A. J., McKeon, B. J., and Marusic, I. (2011). High–Reynolds number wall turbulence. *Annual Review of Fluid Mechanics*, 43(1):353–375.

Spalart, P. R. (1988). Direct simulation of a turbulent boundary layer up to $R_\theta = 1410$. *Journal of Fluid Mechanics*, 187:61–98.

Swearingen, J. D. and Blackwelder, R. F. (1987). The growth and breakdown of streamwise vortices in the presence of a wall. *Journal of Fluid Mechanics*, 182:255–290.

Talamelli, A., Persiani, F., H M Fransson, J., Alfredsson, P. H., V Johansson, A., M Nagib, H., Rüedi, J.-D., R Sreenivasan, K., and A Monkewitz, P. (2009). CICLoPE—a response to the need for high Reynolds number experiments. *Fluid Dynamics Research*, 41(2):021407.

Taylor, G. I. (1935). Statistical theory of turbulence. *Proceedings of the Royal Society of London. Series A - Mathematical and Physical Sciences*, 151(873):421–444.

Taylor, G. I. (1938). The spectrum of turbulence. *Proceedings of the Royal Society of London. Series A - Mathematical and Physical Sciences*, 164(919):476–490.

Tennekes, H. and Lumley, J. L. (1972). *A First Course in Turbulence*. MIT press.

Tennekes, H. and Wyngaard, J. C. (1972). The intermittent small-scale structure of turbulence: Data-processing hazards. *Journal of Fluid Mechanics*, 55:93–103.

Theodorsen, T. (1952). Mechanism of turbulence. In *Proc. of the 2nd Midwestern Conference on Fluid Mechanics*, Ohio State University.

Townsend, A. A. (1976). *The Structure of Turbulent Shear Flow*. Cambridge University Press, second edition.

Uhlmann, M., Pinelli, A., Kawahara, G., and Sekimoto, A. (2007). Marginally turbulent flow in a square duct. *Journal of Fluid Mechanics*, 588:153–162.

Unger, F. (1994). *Numerische Simulation turbulenter Rohrströmungen*. Dissertation, Technische Universität München.

Vinokur, M. (1983). On one-dimensional stretching functions for finite-difference calculations. *Journal of Computational Physics*, 50(2):215–234.

von Kameke, A., Huhn, F., Fernández-Garćia, G., Muñuzuri, A. P., and Pérez-Muñuzuri, V. (2011). Double cascade turbulence and Richardson dispersion in a horizontal fluid flow induced by Faraday waves. *Physical Review Letters*, 107(7):074502.

von Kármán, T. (1930). Mechanische Änlichkeit und Turbulenz. *Nachrichten von der Gesellschaft der Wissenschaften zu Göttingen, Mathematisch-Physikalische Klasse*, 1930:58–76.

Wagner, C. (1995). *Direkte numerische Simulation turbulenter Strömungen in einer Rohrerweiterung*. Dissertation, Technische Universität München.

Wagner, C. and Friedrich, R. (1998). On the turbulence structure in solid and permeable pipes. *International Journal of Heat and Fluid Flow*, 19(5):459–469.

Wagner, C., Hüttl, T. J., and Friedrich, R. (2001). Low-Reynolds-number effects derived from direct numerical simulations of turbulent pipe flow. *Computers and Fluids*, 30:581–590.

Wagner, S. and Shishkina, O. (2013). Aspect-ratio dependency of Rayleigh–Bénard convection in box-shaped containers. *Physics of Fluids*, 25(8):085110.

Wallace, J. M. (2016). Quadrant analysis in turbulence research: History and evolution. *Annual Review of Fluid Mechanics*, 48(1):131–158.

Wallace, J. M., Eckelmann, H., and Brodkey, R. S. (1972). The wall region in turbulent shear flow. *Journal of Fluid Mechanics*, 54(1):39–48.

Wang, G. and Richter, D. H. (2019). Two mechanisms of modulation of very-large-scale motions by inertial particles in open channel flow. *Journal of Fluid Mechanics*, 868:538–559.

Westerweel, J., Adrian, R. J., Eggels, J. G. M., and Nieuwstadt, F. T. M. (1993). Measurements with particle image velocimetry on fully developed turbulent pipe flow at low Reynolds number. In Adrian, R. J., Durão, D. F. G., Durst, F., Heitor, M. V., Maeda, M., and Whitelaw, J. H., editors, *Laser Techniques and Applications in Fluid Mechanics*, pages 285–307. Springer Berlin Heidelberg, Berlin, Heidelberg.

Westerweel, J., Draad, A. A., van der Hoeven, J. G. T., and van Oord, J. (1996). Measurement of fully-developed turbulent pipe flow with digital particle image velocimetry. *Experiments in Fluids*, 20(3):165–177.

Willert, C. E., Soria, J., Stanislas, M., Klinner, J., Amili, O., Eisfelder, M., Cuvier, C., Bellani, G., Fiorini, T., and Talamelli, A. (2017). Near-wall statistics of a turbulent pipe flow at shear Reynolds numbers up to 40 000. *Journal of Fluid Mechanics*, 826:R5.

Willmarth, W. W. and Lu, S. S. (1972). Structure of the reynolds stress near the wall. *Journal of Fluid Mechanics*, 55(1):65–92.

Wu, X., Baltzer, J. R., and Adrian, R. J. (2012). Direct numerical simulation of a 30R long turbulent pipe flow at R^+=685: Large- and very large-scale motions. *Journal of Fluid Mechanics*, 698:235–281.

Wu, X. and Moin, P. (2008). A direct numerical simulation study on the mean velocity characteristics in turbulent pipe flow. *Journal of Fluid Mechanics*, 608:81–112.

Xu, C., Zhang, Z., den Toonder, J. M. J., and Nieuwstadt, F. T. M. (1996). Origin of high kurtosis levels in the viscous sublayer. Direct numerical simulation and experiment. *Physics of Fluids*, 8(7):1938–1944.

Yamamoto, Y. and Tsuji, Y. (2018). Numerical evidence of logarithmic regions in channel flow at Re_τ=8000. *Physical Review Fluids*, 3(1):012602.

Zagarola, M. V., Perry, A. E., and Smits, A. J. (1997). Log laws or power laws: The scaling in the overlap region. *Physics of Fluids*, 9(7):2094–2100.

Zagarola, M. V. and Smits, A. J. (1997). Scaling of the mean velocity profile for turbulent pipe flow. *Physical Review Letters*, 78(2):239–242.

Zagarola, M. V. and Smits, A. J. (1998). Mean-flow scaling of turbulent pipe flow. *Journal of Fluid Mechanics*, 373:33–79.

Zimmerman, S., Philip, J., Monty, J., Talamelli, A., Marusic, I., Ganapathisubramani, B., Hearst, R. J., Bellani, G., Baidya, R., Samie, M., Zheng, X., Dogan, E., Mascotelli, L., and Klewicki, J. (2019). A comparative study of the velocity and vorticity structure in pipes and boundary layers at friction Reynolds numbers up to 10^4. *Journal of Fluid Mechanics*, 869:182–213.

Appendix A

Mathematical Relations and Operators

In the following, relations between the Cartesian and the cylindrical coordinate system as well as mathematical operators used throughout this thesis are defined.

Relation between the Cartesian and the cylindrical coordinate system

The relation between the base the base $\{\mathbf{e}_x, \mathbf{e}_y, \mathbf{e}_z\}$ of a Cartesian coordinate system (x, y, z) and the base $\{\mathbf{e}_r, \mathbf{e}_\varphi, \mathbf{e}_z\}$ of a cylindrical coordinate system (r, φ, z) is given as follows

$$\mathbf{e}_r = \cos(\varphi)\mathbf{e}_x + \sin(\varphi)\mathbf{e}_y \tag{A.1}$$

$$\mathbf{e}_\varphi = -\sin(\varphi)\mathbf{e}_x + \cos(\varphi)\mathbf{e}_y \tag{A.2}$$

$$\mathbf{e}_z = \mathbf{e}_z \tag{A.3}$$

The Cartesian position vector $\mathbf{x} = (x, y, z)^T$ can be expressed by cylindrical coordinates

$$\mathbf{x} = \begin{pmatrix} x \\ y \\ z \end{pmatrix} = \begin{pmatrix} r\cos(\varphi) \\ r\sin(\varphi) \\ z \end{pmatrix}, \tag{A.4}$$

and the cylindrical position vector $\mathbf{r} = (r, \varphi, z)^T$ can be expressed by Cartesian coordinates

$$\mathbf{r} = \begin{pmatrix} r \\ \varphi \\ z \end{pmatrix} = \begin{pmatrix} \sqrt{x^2 + y^2} \\ \arctan(y/x) \\ z \end{pmatrix}. \tag{A.5}$$

The velocity $\mathbf{v}$ is defined as the rate of change of the location $\mathbf{x}$

$$\begin{aligned} \mathbf{v} &= u\mathbf{e}_x + v\mathbf{e}_y + w\mathbf{e}_z \\ &= \frac{\partial}{\partial t}(x\mathbf{e}_x + y\mathbf{e}_y + z\mathbf{e}_z) \\ &= \frac{\partial}{\partial t}(r\cos(\varphi)\mathbf{e}_x + r\sin(\varphi)\mathbf{e}_y + z\mathbf{e}_z) \\ &= (\dot{r}\cos(\varphi) - r\dot{\varphi}\sin(\varphi))\,\mathbf{e}_x + (\dot{r}\sin(\varphi) + r\dot{\varphi}\cos(\varphi))\,\mathbf{e}_y + \dot{z}\mathbf{e}_z \\ &= \dot{r}\mathbf{e}_r + r\dot{\varphi}\mathbf{e}_\varphi + \dot{z}\mathbf{e}_z \\ &= u_r\mathbf{e}_r + u_\varphi\mathbf{e}_\varphi + u_z\mathbf{e}_z \end{aligned} \tag{A.6}$$

Here we use the notation $\hat{a} = \partial a/\partial t$ for the rate of change of a.

Mathematical Operators

The scalar product between two three-dimensional vectors $\mathbf{a} = (a_1, a_2, a_3)^T$ and $\mathbf{b} = (b_1, b_2, b_3)^T$ is defined as

$$\mathbf{a} \cdot \mathbf{b} = a_1b_1 + a_2b_2 + a_3b_3, \tag{A.7}$$

and the dyadic product between $\mathbf{a}$ and $\mathbf{b}$ is defined as

$$\mathbf{a} \otimes \mathbf{b} = \begin{pmatrix} a_1b_1 & a_1b_2 & a_1b_3 \\ a_2b_1 & a_2b_2 & a_2b_3 \\ a_3b_1 & a_3b_2 & a_3b_3 \end{pmatrix}. \tag{A.8}$$

The Nabla operator ∇ is a vector differential operator. Applying the Nabla operator to a scalar a, a vector $\mathbf{a}$, and a tensor A in Cartesian coordinates results in

$$\nabla a = \frac{\partial a}{\partial x}\mathbf{e}_x + \frac{\partial a}{\partial y}\mathbf{e}_y + \frac{\partial a}{\partial z}\mathbf{e}_z, \tag{A.9}$$

$$\nabla \cdot \mathbf{a} = \frac{\partial a_x}{\partial x} + \frac{\partial a_y}{\partial y} + \frac{\partial a_z}{\partial z}, \tag{A.10}$$

$$\begin{aligned} \nabla \cdot A = & \left(\frac{\partial A_{xx}}{\partial x} + \frac{\partial A_{yx}}{\partial y} + \frac{\partial A_{zx}}{\partial z}\right)\mathbf{e}_x \\ & + \left(\frac{\partial A_{xy}}{\partial x} + \frac{\partial A_{yy}}{\partial y} + \frac{\partial A_{zy}}{\partial z}\right)\mathbf{e}_y \\ & + \left(\frac{\partial A_{xz}}{\partial x} + \frac{\partial A_{yz}}{\partial y} + \frac{\partial A_{zz}}{\partial z}\right)\mathbf{e}_z. \end{aligned} \tag{A.11}$$

In the cylindrical coordinate system the application of the Nabla operator yields

$$\nabla a = \frac{\partial a}{\partial r}\mathbf{e}_r + \frac{1}{r}\frac{\partial a}{\partial \varphi}\mathbf{e}_\varphi + \frac{\partial a}{\partial z}\mathbf{e}_z, \tag{A.12}$$

$$\nabla \cdot \mathbf{a} = \frac{1}{r}\frac{\partial (r a_r)}{\partial r} + \frac{1}{r}\frac{\partial a_\varphi}{\partial \varphi} + \frac{\partial a_z}{\partial z}, \tag{A.13}$$

$$\begin{aligned} \nabla \cdot A = & \left(\frac{1}{r}\frac{\partial (r A_{rr})}{\partial r} + \frac{1}{r}\frac{\partial A_{\varphi r}}{\partial \varphi} + \frac{\partial A_{zr}}{\partial z} - \frac{A_{\varphi\varphi}}{r}\right)\mathbf{e}_r \\ & + \left(\frac{1}{r}\frac{\partial (r A_{r\varphi})}{\partial r} + \frac{1}{r}\frac{\partial A_{\varphi\varphi}}{\partial \varphi} + \frac{\partial A_{z\varphi}}{\partial z} + \frac{A_{r\varphi}}{r}\right)\mathbf{e}_\varphi \\ & + \left(\frac{1}{r}\frac{(r \partial A_{rz})}{\partial r} + \frac{1}{r}\frac{\partial A_{\varphi z}}{\partial \varphi} + \frac{\partial A_{zz}}{\partial z}\right)\mathbf{e}_z. \end{aligned} \tag{A.14}$$

Appendix B

Derivation of the Turbulent Kinetic Energy Budget in Cylindrical Coordinates

Applying the averaging operator to the momentum equations in cylindrical coordinates (1.10b)–(1.10d), while taking into account fully-developed turbulent pipe flow with two homogeneous directions, results in the Reynolds averaged equations in cylindrical coordinates

$$\frac{1}{r}\frac{\partial(r\langle u_r' u_z'\rangle)}{\partial r} + \frac{\partial\langle p\rangle}{\partial z} = \frac{1}{\mathrm{Re}_\tau}\frac{1}{r}\frac{\partial}{\partial r}\left(r\frac{\partial\langle u_z\rangle}{\partial r}\right), \tag{B.1a}$$

$$\frac{\partial\langle u_r' u_r'\rangle}{\partial r} + \frac{1}{r}\left(\langle u_r' u_r'\rangle - \langle u_\varphi' u_\varphi'\rangle\right) + \frac{\partial\langle p\rangle}{\partial r} = 0. \tag{B.1b}$$

Note that in the azimuthal equation all average terms equal to zero. Subtracting equation (B.1a) from the axial momentum equation (1.10b) and applying the Reynolds decomposition yields a transport equation for the fluctuating axial velocity

$$\frac{\partial u_z'}{\partial t} + (\langle u_z\rangle + u_z')\frac{\partial u_z'}{\partial z} + \frac{1}{r}u_\varphi'\frac{\partial u_z'}{\partial\varphi} + u_r'\frac{\partial(\langle u_z\rangle + u_z')}{\partial r} - \frac{1}{r}\frac{\partial(r\langle u_r' u_z'\rangle)}{\partial r} + \frac{\partial p'}{\partial z} = \frac{1}{\mathrm{Re}_\tau}\nabla^2 u_z' \tag{B.2}$$

Equation (B.2) is then multiplied by u_z' resulting in

$$\begin{aligned} u_z'\frac{\partial u_z'}{\partial t} + u_z'(\langle u_z\rangle + u_z')\frac{\partial u_z'}{\partial z} &+ u_z'\frac{1}{r}u_\varphi'\frac{\partial u_z'}{\partial\varphi} + u_z' u_r'\frac{\partial(\langle u_z\rangle + u_z')}{\partial r} \\ &- u_z'\frac{1}{r}\frac{\partial(r\langle u_r' u_z'\rangle)}{\partial r} + u_z'\frac{\partial p'}{\partial z} = u_z'\frac{1}{\mathrm{Re}_\tau}\nabla^2 u_z', \end{aligned} \tag{B.3}$$

where the left hand side can be elaborated as follows

$$
\begin{aligned}
& u_z'\frac{\partial u_z'}{\partial t} + u_z'(\langle u_z\rangle + u_z')\frac{\partial u_z'}{\partial z} + u_z'\frac{1}{r}u_\varphi'\frac{\partial u_z'}{\partial \varphi} + u_z'u_r'\frac{\partial(\langle u_z\rangle + u_z')}{\partial r} - u_z'\frac{1}{r}\frac{\partial(r\langle u_r'u_z'\rangle)}{\partial r} + u_z'\frac{\partial p'}{\partial z} \\
= & \frac{1}{2}\frac{\partial(u_z'u_z')}{\partial t} + (\langle u_z\rangle + u_z')\frac{1}{2}\frac{\partial(u_z'u_z')}{\partial z} + \frac{1}{2r}u_\varphi'\frac{\partial(u_z'u_z')}{\partial\varphi} + u_z'u_r'\frac{\langle u_z\rangle}{\partial r} + \frac{1}{2}u_r'\frac{\partial(u_z'u_z')}{\partial r} \\
& - u_z'\frac{1}{r}\frac{\partial(r\langle u_r'u_z'\rangle)}{\partial r} + u_z'\frac{\partial p'}{\partial z} \\
= & \frac{1}{2}\frac{\partial(u_z'u_z')}{\partial t} + \langle u_z\rangle\frac{1}{2}\frac{\partial(u_u'u_z')}{\partial z} + \frac{1}{2}\frac{\partial(u_z'u_z'u_z')}{\partial z} + \frac{1}{2r}\frac{\partial(u_\varphi'u_z'u_z')}{\partial\varphi} + \frac{1}{2}\frac{\partial(u_r'u_z'u_z')}{\partial r} + u_z'u_r'\frac{\langle u_z\rangle}{\partial r} \\
& -\frac{1}{2}u_z'u_z'\underbrace{\left(\frac{\partial u_z'}{\partial z} + \frac{1}{r}\frac{\partial u_\varphi'}{\varphi} + \frac{\partial u_r'}{\partial r}\right)}_{=-u_r'/r\ (1.10a)} - u_z'\frac{1}{r}\frac{\partial(r\langle u_r'u_z'\rangle)}{\partial r} + \frac{\partial(p'u_z')}{\partial z} - p'\frac{\partial u_z'}{\partial z},
\end{aligned}
\tag{B.4}
$$

and the right hand side as follows

$$
\begin{aligned}
u_z'\frac{1}{\mathrm{Re}_\tau}\nabla^2 u_z' &= \frac{1}{\mathrm{Re}_\tau}\left[u_z'\frac{\partial^2 u_z'}{\partial z^2} + u_z'\frac{1}{r^2}\frac{\partial^2 u_z}{\partial\varphi^2} + u_z'\frac{1}{r}\frac{\partial}{\partial r}\left(r\frac{\partial u_z'}{\partial r}\right)\right] \\
&= \frac{1}{\mathrm{Re}_\tau}\left[\frac{\partial}{\partial z}\left(u_z'\frac{\partial u_z'}{\partial z}\right) - \left(\frac{\partial u_z'}{\partial z}\right)^2 + \frac{1}{r^2}\frac{\partial}{\partial\varphi}\left(u_z'\frac{\partial u_z'}{\partial\varphi}\right) - \frac{1}{r^2}\left(\frac{\partial u_z'}{\partial\varphi}\right)^2\right. \\
&\quad \left. + \frac{\partial}{\partial r}\left(u_z'\frac{\partial u_z'}{\partial r}\right) - \left(\frac{\partial u_z'}{\partial r}\right)^2 + \frac{1}{2r}\frac{\partial(u_z'u_z')}{\partial r}\right] \\
&= \frac{1}{\mathrm{Re}_\tau}\left[\frac{1}{2}\frac{\partial^2(u_z'u_z')}{\partial z^2} - \left(\frac{\partial u_z'}{\partial z}\right)^2 + \frac{1}{2r^2}\frac{\partial^2(u_z'u_z')}{\partial\varphi^2} - \frac{1}{r^2}\left(\frac{\partial u_z'}{\partial\varphi}\right)^2\right. \\
&\quad \left. + \frac{1}{2}\frac{\partial^2(u_z'u_z')}{\partial r^2} - \left(\frac{\partial u_z'}{\partial r}\right)^2 + \frac{1}{2r}\frac{\partial(u_z'u_z')}{\partial r}\right].
\end{aligned}
\tag{B.5}
$$

Substituting expressions (B.4) and (B.5) back into equation (B.3) and taking the average yields

$$
\begin{aligned}
& \frac{1}{2r}\frac{\partial(r\langle u_r'u_z'u_z'\rangle)}{\partial r} + \langle u_z'u_r'\rangle\frac{\langle u_z\rangle}{\partial r} - \left\langle p'\frac{\partial u_z'}{\partial z}\right\rangle \\
= & \frac{1}{2\mathrm{Re}_\tau}\left[\frac{1}{r}\frac{\partial}{\partial r}\left(r\frac{\partial\langle u_z'u_z'\rangle}{\partial r}\right)\right] - \frac{1}{\mathrm{Re}_\tau}\left[\left\langle\left(\frac{\partial u_z'}{\partial z}\right)^2\right\rangle + \frac{1}{r^2}\left\langle\left(\frac{\partial u_z'}{\partial\varphi}\right)^2\right\rangle + \left\langle\left(\frac{\partial u_z'}{\partial r}\right)^2\right\rangle\right]
\end{aligned}
\tag{B.6}
$$

which is a transport equation for the axial component of the turbulent kinetic energy — or in other words the streamwise Reynolds stress budget divided by two. Next, inserting the Reynolds decomposition to the azimuthal momentum equation results in

$$\begin{aligned} \frac{\partial u'_\varphi}{\partial t} &+ (\langle u_z\rangle + u'_z)\frac{\partial u'_\varphi}{\partial z} + \frac{1}{r}\left(u'_\varphi\frac{\partial u'_\varphi}{\partial \varphi} + u'_r u'_\varphi\right) + u'_r\frac{\partial u'_\varphi}{\partial r} + \frac{1}{r}\frac{\partial p'}{\partial \varphi} \\ &= \frac{1}{\mathrm{Re}_\tau}\left[\nabla^2 u'_\varphi - \frac{1}{r^2}\left(u'_\varphi - 2\frac{\partial u'_r}{\partial \varphi}\right)\right]. \end{aligned} \tag{B.7}$$

Equation (B.7) is now multiplied by u'_φ yielding

$$\begin{aligned} u'_\varphi\frac{\partial u'_\varphi}{\partial t} &+ u'_\varphi\langle u_z\rangle\frac{\partial u'_\varphi}{\partial z} + u'_\varphi u'_z\frac{\partial u'_\varphi}{\partial z} + \frac{1}{r}u'_\varphi u'_\varphi\frac{\partial u'_\varphi}{\partial \varphi} + \frac{u'_r u'_\varphi u'_\varphi}{r} + u'_\varphi u'_r\frac{\partial u'_\varphi}{\partial r} \\ &+ \frac{1}{r}u'_\varphi\frac{\partial p'}{\partial \varphi} = u'_\varphi\frac{1}{\mathrm{Re}_\tau}\left[\nabla^2 u'_\varphi - \frac{1}{r^2}\left(u'_\varphi - 2\frac{\partial u'_r}{\partial \varphi}\right)\right]. \end{aligned} \tag{B.8}$$

Next, we rearrange both the left hand side

$$\begin{aligned} & u'_\varphi\frac{\partial u'_\varphi}{\partial t} + u'_\varphi\langle u_z\rangle\frac{\partial u'_\varphi}{\partial z} + u'_\varphi u'_z\frac{\partial u'_\varphi}{\partial z} + \frac{1}{r}u'_\varphi u'_\varphi\frac{\partial u'_\varphi}{\partial \varphi} + u'_\varphi u'_r\frac{\partial u'_\varphi}{\partial r} + \frac{u'_r u'_\varphi u'_\varphi}{r} \\ & + u'_\varphi u'_r\frac{\partial u'_\varphi}{\partial r} + \frac{1}{r}u'_\varphi\frac{\partial p'}{\partial \varphi} \\ = {} & \frac{1}{2}\frac{\partial(u'_\varphi u'_\varphi)}{\partial t} + \frac{1}{2}\langle u_z\rangle\frac{\partial(u'_\varphi u'_\varphi)}{\partial z} + \frac{1}{2}u'_z\frac{\partial(u'_\varphi u'_\varphi)}{\partial z} + \frac{1}{2r}u'_\varphi\frac{\partial(u'_\varphi u'_\varphi)}{\partial \varphi} + \frac{u'_r u'_\varphi u'_\varphi}{r} \\ & + \frac{1}{2}u'_r\frac{\partial(u'_\varphi u'_\varphi)}{\partial r} + \frac{1}{r}u'_\varphi\frac{\partial p'}{\partial \varphi} \\ = {} & \frac{1}{2}\frac{\partial(u'_\varphi u'_\varphi)}{\partial t} + \frac{1}{2}\langle u_z\rangle\frac{\partial(u'_\varphi u'_\varphi)}{\partial z} + \frac{1}{2}\frac{\partial(u'_z u'_\varphi u'_\varphi)}{\partial z} + \frac{1}{2r}\frac{\partial(u'_\varphi u'_\varphi u'_\varphi)}{\partial \varphi} + \frac{u'_r u'_\varphi u'_\varphi}{r} \\ & + \frac{1}{2}\frac{\partial(u'_r u'_\varphi u'_\varphi)}{\partial r} - \frac{1}{2}u'_\varphi u'_\varphi\underbrace{\left(\frac{\partial u'_z}{\partial z} + \frac{1}{r}\frac{\partial u'_\varphi}{\varphi} + \frac{\partial u'_r}{\partial r}\right)}_{=-u'_r/r\ (1.10a)} + \frac{1}{r}\frac{\partial(u'_\varphi p')}{\partial \varphi} - \frac{1}{r}p'\frac{\partial u'_\varphi}{\partial \varphi}, \end{aligned} \tag{B.9}$$

and the right hand side of equation (B.8),

$$\begin{aligned}
& u'_\varphi \frac{1}{\mathrm{Re}_\tau}\left[\nabla^2 u'_\varphi - \frac{1}{r^2}\left(u'_\varphi - 2\frac{\partial u'_r}{\partial \varphi}\right)\right] \\
= & \frac{1}{\mathrm{Re}_\tau}\left[u'_\varphi \frac{\partial^2 u'_\varphi}{\partial z^2} + u'_\varphi \frac{1}{r^2}\frac{\partial^2 u'_\varphi}{\partial \varphi^2} + u'_\varphi \frac{1}{r}\frac{\partial}{\partial r}\left(r\frac{\partial u'_\varphi}{\partial r}\right) - \frac{u'_\varphi u'_\varphi}{r^2} + \frac{2}{r^2}u'_\varphi \frac{\partial u'_r}{\partial \varphi}\right] \\
= & \frac{1}{\mathrm{Re}_\tau}\left[\frac{\partial}{\partial z}\left(u'_\varphi \frac{\partial u'_\varphi}{\partial z}\right) - \left(\frac{\partial u'_\varphi}{\partial z}\right)^2 + \frac{1}{r^2}\frac{\partial}{\partial \varphi}\left(u'_\varphi \frac{\partial u'_\varphi}{\partial \varphi}\right) - \frac{1}{r^2}\left(\frac{\partial u'_\varphi}{\partial \varphi}\right)^2\right. \\
& \left. + \frac{\partial}{\partial r}\left(u'_\varphi \frac{\partial u'_\varphi}{\partial r}\right) - \left(\frac{\partial u'_\varphi}{\partial r}\right)^2 + \frac{1}{2r}\frac{\partial (u'_\varphi u'_\varphi)}{\partial r} - \frac{u'_\varphi u'_\varphi}{r^2} + \frac{2}{r^2}u'_\varphi \frac{\partial u'_r}{\partial \varphi}\right] \\
= & \frac{1}{\mathrm{Re}_\tau}\left[\frac{1}{2}\frac{\partial^2 (u'_\varphi u'_\varphi)}{\partial z^2} - \left(\frac{\partial u'_\varphi}{\partial z}\right)^2 + \frac{1}{2r^2}\frac{\partial^2 (u'_\varphi u'_\varphi)}{\partial \varphi^2} - \frac{1}{r^2}\left(\frac{\partial u'_\varphi}{\partial \varphi}\right)^2\right. \\
& \left. + \frac{1}{2}\frac{\partial^2 (u'_\varphi u'_\varphi)}{\partial r^2} - \left(\frac{\partial u'_\varphi}{\partial r}\right)^2 + \frac{1}{2r}\frac{\partial (u'_\varphi u'_\varphi)}{\partial r} - \frac{u'_\varphi u'_\varphi}{r^2} + \frac{2}{r^2}u'_\varphi \frac{\partial u'_r}{\partial \varphi}\right].
\end{aligned} \tag{B.10}$$

We substitute expressions (B.9) and (B.10) back into equation (B.8) and apply the averaging operation to obtain the azimuthal component of turbulent kinetic energy budget

$$\begin{aligned}
& \frac{1}{2r}\frac{\partial (r\langle u'_r u'_\varphi u'_\varphi\rangle)}{\partial r} + \frac{\langle u'_r u'_\varphi u'_\varphi\rangle}{r} - \frac{1}{r}\left\langle p'\frac{\partial u'_\varphi}{\partial \varphi}\right\rangle \\
= & \frac{1}{2\mathrm{Re}_\tau}\left[\frac{1}{r}\frac{\partial}{\partial r}\left(r\frac{\partial \langle u'_\varphi u'_\varphi\rangle}{\partial r}\right) - \frac{2}{r^2}(\langle u'_\varphi u'_\varphi\rangle - \langle u'_r u'_r\rangle)\right] \\
& - \frac{1}{\mathrm{Re}_\tau}\left[\left\langle\left(\frac{\partial u'_\varphi}{\partial z}\right)^2\right\rangle + \frac{1}{r^2}\left\langle\left(\frac{\partial u'_\varphi}{\partial \varphi} + u'_r\right)^2\right\rangle + \left\langle\left(\frac{\partial u'_\varphi}{\partial r}\right)^2\right\rangle\right]
\end{aligned} \tag{B.11}$$

which corresponds to the azimuthal streamwise Reynolds stress budget divided by two. Next, the radial Reynolds equation (B.1b) is subtracted from the radial momentum equation (1.10d) and the Reynolds decomposition is applied, which yields

$$\begin{aligned}
\frac{\partial u'_r}{\partial t} & + (\langle u_z\rangle + u'_z)\frac{\partial u'_r}{\partial z} + \frac{1}{r}\left(u'_\varphi \frac{\partial u'_r}{\partial \varphi} - u'_\varphi u'_\varphi\right) + u'_r\frac{\partial u'_r}{\partial r} + \frac{\partial p'}{\partial r} - \frac{\partial \langle u'_r u'_r\rangle}{\partial r} \\
& - \frac{1}{r}\left(\langle u'_r u'_r\rangle - \langle u'_\varphi u'_\varphi\rangle\right) = \frac{1}{\mathrm{Re}_\tau}\left[\nabla^2 u'_r - \frac{1}{r^2}\left(u'_r + 2\frac{\partial u'_\varphi}{\partial \varphi}\right)\right].
\end{aligned} \tag{B.12}$$

Multiplying equation (B.12) with u'_r gives

$$\begin{aligned} u'_r\frac{\partial u'_r}{\partial t} &+ u'_r\langle u_z\rangle\frac{\partial u'_r}{\partial z} + u'_r u'_z\frac{\partial u'_r}{\partial z} + \frac{1}{r}u'_r u'_\varphi\frac{\partial u'_r}{\partial \varphi} - \frac{u'_r u'_\varphi u'_\varphi}{r} + u'_r u'_r\frac{\partial u'_r}{\partial r} + u'_r\frac{\partial p'}{\partial r} \\ &- u'_r\frac{\partial\langle u'_r u'_r\rangle}{\partial r} - u'_r\frac{1}{r}\left(\langle u'_r u'_r\rangle - \langle u'_\varphi u'_\varphi\rangle\right) = u'_r\frac{1}{\mathrm{Re}_\tau}\left[\nabla^2 u'_r - \frac{1}{r^2}\left(u'_r + 2\frac{\partial u'_\varphi}{\partial \varphi}\right)\right] \end{aligned} \quad (B.13)$$

The left hand side of equation (B.13) can be elaborated to

$$\begin{aligned} & u'_r\frac{\partial u'_r}{\partial t} + u'_r\langle u_z\rangle\frac{\partial u'_r}{\partial z} + u'_r u'_z\frac{\partial u'_r}{\partial z} + \frac{1}{r}u'_r u'_\varphi\frac{\partial u'_r}{\partial \varphi} - \frac{u'_r u'_\varphi u'_\varphi}{r} + u'_r u'_r\frac{\partial u'_r}{\partial r} + u'_r\frac{\partial p'}{\partial r} \\ & -u'_r\frac{\partial\langle u'_r u'_r\rangle}{\partial r} - u'_r\frac{1}{r}\left(\langle u'_r u'_r\rangle - \langle u'_\varphi u'_\varphi\rangle\right) \\ = \; & \frac{1}{2}\frac{\partial(u'_r u'_r)}{\partial t} + \frac{1}{2}\langle u_z\rangle\frac{\partial(u'_r u'_r)}{\partial z} + \frac{1}{2}u'_z\frac{\partial(u'_r u'_r)}{\partial z} + \frac{1}{2r}u'_\varphi\frac{\partial(u'_r u'_r)}{\partial \varphi} - \frac{u'_r u'_\varphi u'_\varphi}{r} + \frac{1}{2}u'_r\frac{\partial(u'_r u'_r)}{\partial r} \\ & +u'_r\frac{\partial p'}{\partial r} - u'_r\frac{\partial\langle u'_r u'_r\rangle}{\partial r} - u'_r\frac{1}{r}\left(\langle u'_r u'_r\rangle - \langle u'_\varphi u'_\varphi\rangle\right) \\ = \; & \frac{1}{2}\frac{\partial(u'_r u'_r)}{\partial t} + \frac{1}{2}\langle u_z\rangle\frac{\partial(u'_r u'_r)}{\partial z} + \frac{1}{2}\frac{\partial(u'_z u'_r u'_r)}{\partial z} + \frac{1}{2r}\frac{\partial(u'_\varphi u'_r u'_r)}{\partial \varphi} - \frac{u'_r u'_\varphi u'_\varphi}{r} + \frac{1}{2}\frac{\partial(u'_r u'_r u'_r)}{\partial r} \\ & -\frac{1}{2}u'_r u'_r\underbrace{\left(\frac{\partial u'_z}{\partial z} + \frac{1}{r}\frac{\partial u'_\varphi}{\varphi} + \frac{\partial u'_r}{\partial r}\right)}_{=-u'_r/r \;(1.10a)} + \frac{\partial(p' u'_r)}{\partial r} - p'\frac{\partial u'_r}{\partial r} - u'_r\frac{\partial\langle u'_r u'_r\rangle}{\partial r} - u'_r\frac{1}{r}\left(\langle u'_r u'_r\rangle - \langle u'_\varphi u'_\varphi\rangle\right) \end{aligned} \quad (B.14)$$

and the right hand side yields

$$\begin{aligned} & u'_r\frac{1}{\mathrm{Re}_\tau}\left[\nabla^2 u'_r - \frac{1}{r^2}\left(u'_r + 2\frac{\partial u'_\varphi}{\partial \varphi}\right)\right] \\ = \; & \frac{1}{\mathrm{Re}_\tau}\left[u'_r\frac{\partial^2 u'_r}{\partial z^2} + u'_r\frac{1}{r^2}\frac{\partial^2 u'_r}{\partial \varphi^2} + u'_r\frac{1}{r}\frac{\partial}{\partial r}\left(r\frac{\partial u'_r}{\partial r}\right) - \frac{u'_r u'_r}{r^2} - 2u'_r\frac{\partial u'_\varphi}{\partial \varphi}\right] \\ = \; & \frac{1}{\mathrm{Re}_\tau}\left[\frac{\partial}{\partial z}\left(u'_r\frac{\partial u'_r}{\partial z}\right) - \left(\frac{\partial u'_r}{\partial z}\right)^2 + \frac{1}{r^2}\frac{\partial}{\partial r}\left(u'_r\frac{\partial u'_r}{\partial \varphi}\right) - \frac{1}{r^2}\left(\frac{\partial u'_r}{\partial \varphi}\right)^2\right. \\ & \left. + \frac{\partial}{\partial r}\left(u'_r\frac{\partial u'_r}{\partial r}\right) - \left(\frac{\partial u'_r}{\partial r}\right)^2 + \frac{1}{2r}\frac{\partial(u'_r u'_r)}{\partial r} - \frac{u'_r u'_r}{r^2} - 2u'_r\frac{\partial u'_\varphi}{\partial \varphi}\right] \\ = \; & \frac{1}{\mathrm{Re}_\tau}\left[\frac{1}{2}\frac{\partial^2(u'_r u'_r)}{\partial z^2} - \left(\frac{\partial u'_r}{\partial z}\right)^2 + \frac{1}{2r^2}\frac{\partial^2(u'_r u'_r)}{\partial \varphi^2} - \frac{1}{r^2}\left(\frac{\partial u'_r}{\partial \varphi}\right)^2\right. \\ & \left. + \frac{1}{2}\frac{\partial^2(u'_r u'_r)}{\partial r^2} - \left(\frac{\partial u'_r}{\partial r}\right)^2 + \frac{1}{2r}\frac{\partial(u'_z u'_z)}{\partial r} - \frac{u'_r u'_r}{r^2} - 2u'_r\frac{\partial u'_\varphi}{\partial \varphi}\right] \end{aligned} \quad (B.15)$$

Taking the average of equation (B.13) after being updated with the expressions (B.14) and (B.15) results in the radial component of the turbulent kinetic energy budget

$$
\begin{aligned}
&-\frac{\langle u'_r u'_\varphi u'_\varphi \rangle}{r} + \frac{1}{2r}\frac{\partial (r\langle u'_r u'_r u'_r \rangle)}{\partial r} + \frac{\partial \langle p' u'_r \rangle}{\partial r} - \left\langle p' \frac{\partial u'_r}{\partial r} \right\rangle \\
=\; & \frac{1}{2\mathrm{Re}_\tau}\left[\frac{1}{r}\frac{\partial}{\partial r}\left(r \frac{\partial \langle u'_r u'_r \rangle}{\partial r}\right) - \frac{1}{r^2}(\langle u'_r u'_r \rangle + \langle u'_\varphi u'_\varphi \rangle)\right] \\
& - \frac{1}{\mathrm{Re}_\tau}\left[\left\langle \left(\frac{\partial u'_r}{\partial z}\right)^2 \right\rangle + \frac{1}{r^2}\left\langle \left(\frac{\partial u'_r}{\partial \varphi} - u'_\varphi\right)^2 \right\rangle + \left\langle \left(\frac{\partial u'_r}{\partial r}\right)^2 \right\rangle\right], \qquad \text{(B.16)}
\end{aligned}
$$

or twice the radial Reynolds stress budget, respectively. Finally, taking the sum of the equations (B.6), (B.11), and (B.16) results in the turbulent kinetic energy budget for fully-developed turbulent pipe flow in cylindrical coordinates depicted in equation (1.64).

Appendix C

Domain-length Dependency of the Instantaneous Streamwise Velocity

Complementing the instantaneous snapshots of the fluctuating velocity components shown in figures 3.1–3.6, iso-contours of the instantaneous streamwise velocity u_z are presented here. Note that the azimuthal and radial velocity component are fluctuations, i.e. $u_\varphi = u'_\varphi$ and $u'_r = u_r$ and are thus already included in the figures mentioned above.

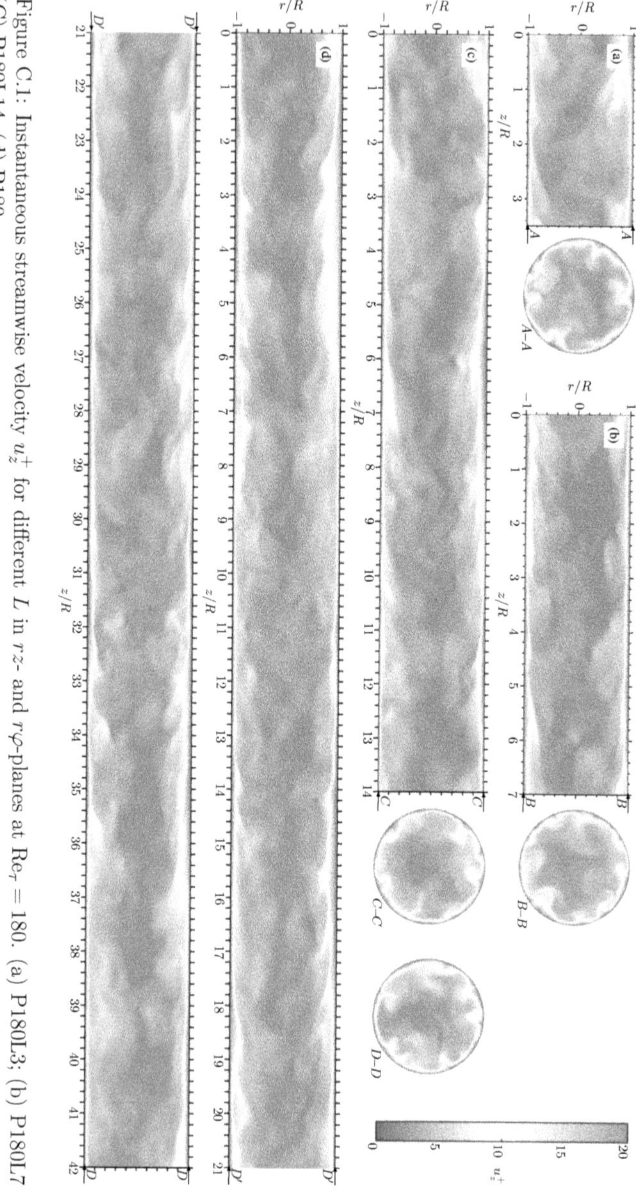

Figure C.1: Instantaneous streamwise velocity u_z^+ for different L in rz- and $r\varphi$-planes at $\mathrm{Re}_\tau = 180$. (a) P180L3; (b) P180L7 (C) P180L14; (d) P180.

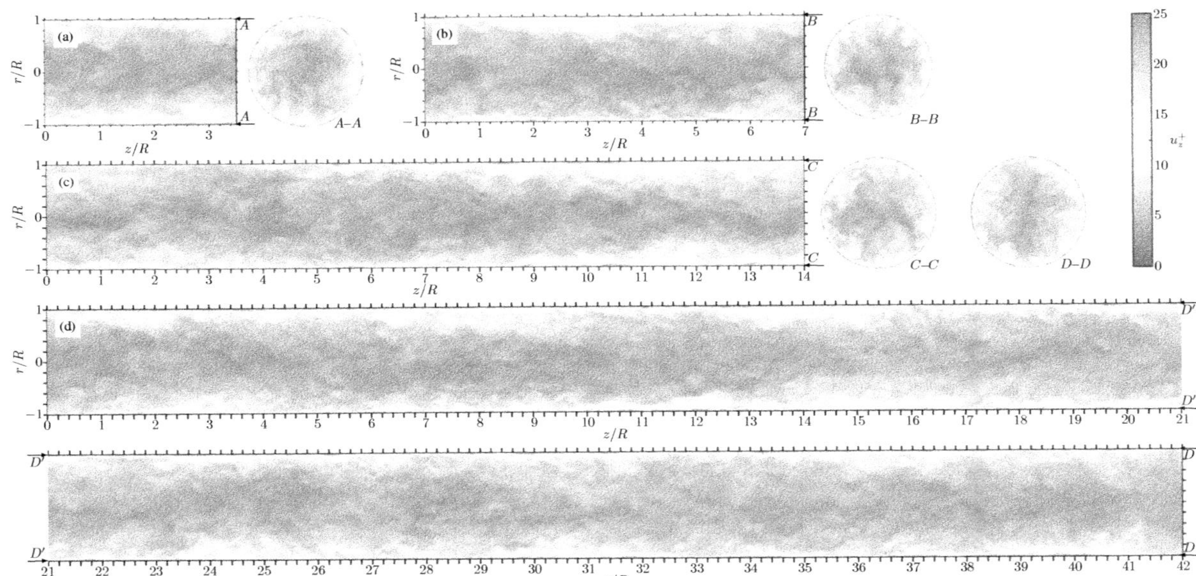

Figure C.2: Instantaneous streamwise velocity u_z^+ for different L in rz- and $r\varphi$-planes at $\mathrm{Re}_\tau = 720$. (a) P720L3; (b) P720L7 (C) P720L14; (d) P720.

Appendix D

Convergence of Statistical Moments in the Bulk Flow

In the bulk region of turbulent pipe flow, statistical quantities are supposed to converge within averaging intervals that scale in bulk units. Figure D.1 shows the turbulence quantities introduced in section 4.2 for P180 in the bulk flow region ($y/R = 0.8$). Here, the averaging coordinate in bulk units reads

$$\Delta\tau_b = \Delta T_b \frac{L}{R} \cdot 2\pi \frac{r}{R} = \Delta T \frac{2\pi u_b L}{R^3}. \tag{D.1}$$

The grey shaded areas indicate a region of $\pm 1\%$ of the final value (see "one-percent criterion" in section 4.2.1).

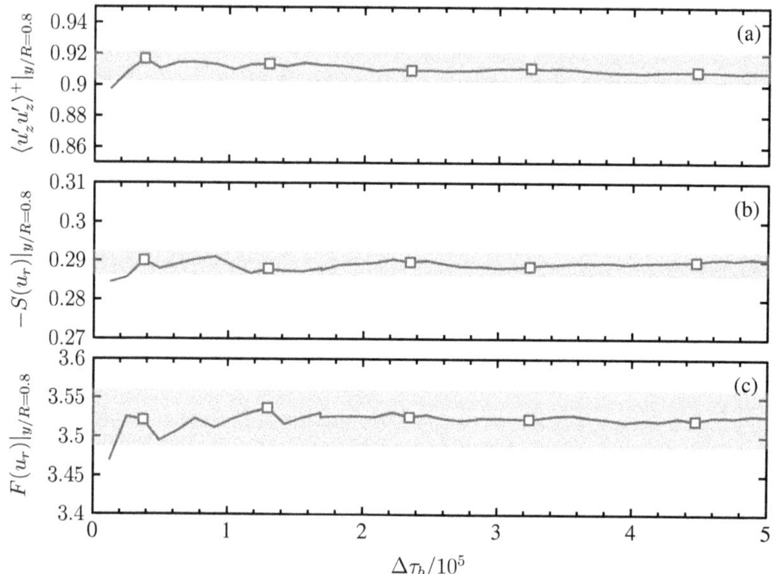

Figure D.1: Convergence of Reynolds stresses and high-order statistical moments over $\Delta\tau_b$ at $y/R = 0.8$ for P180. (a) $\langle u_z' u_z' \rangle^+$; (b) $S(u_r)$; (c) $F(u_r)$.

Appendix E

One-dimensional Pre-multiplied Energy Spectra

To complement the two-dimensional pre-multiplied energy spectra of the streamwise velocity fluctuations — as well as the iso-contours of one-dimensional spectra in the wall-normal-coordinate wave number space — presented in the chapters 3 and 4, the corresponding profiles of the one-dimensional spectra are presented here. The one-dimensional spectra are easily obtained from two-dimensional spectra by integrating over one of the wave number coordinates. Figure E.1 shows one-dimensional pre-multiplied energy spectra of the streamwise velocity fluctuation from different Reynolds numbers at the same wall-normal locations in wall units. Note that unlike in the main part, here the spectra for $\mathrm{Re}_\tau = 2880$, which are far from being converged [21], are included. From the wall distance $y^+ = 15$ the spectral peak locations ($\lambda_z^+ \approx 1000$, $\lambda_\varphi^+ \approx 120$) correspond to the most energetic species of coherent motions, wall-layer streaks. When moving away from the wall, energy is removed from the wavemodes corresponding to wall-layer streaks and larger wavelengths become energetic. For larger Reynolds numbers $\mathrm{Re}_\tau \gtrsim 720$, energy related to outer-flow VLSMs can be found in the spectra even close to wall — although those energies are little when compared to the local energy peak. For these larger Reynolds numbers, the spectra tend towards a κ^{-1} range in the logarithmic layer ($y^+ \gtrsim 30, y/R \lesssim 0.3$). For the lowest Reynolds number ($\mathrm{Re}_\tau = 180$), on the contrary, the narrow range of scales does not allow a second dominant wavemode to develop and the spectra feature a clear one-peak state at all distances from the wall. Figure E.2 shows one-dimensional pre-multiplied energy spectra of the streamwise velocity fluctuation from different Reynolds numbers at the

[21] Two-dimensional spectra take much more samples to converge than one-dimensional spectra, since one averaging coordinate is missing with respect to the latter.

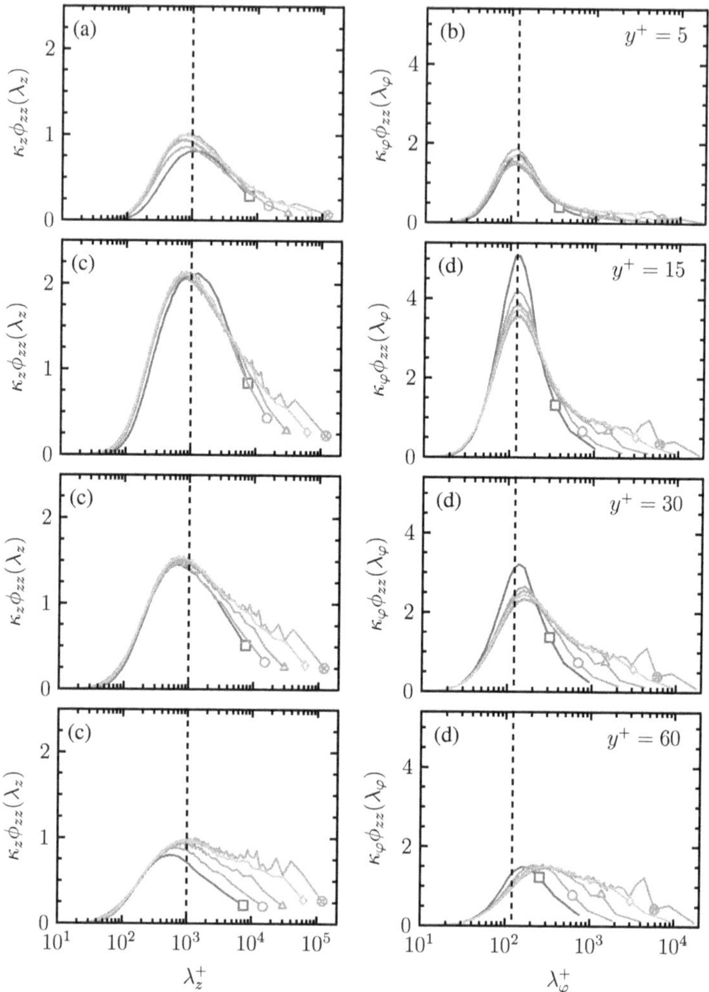

Figure E.1: One-dimensional pre-multiplied energy spectra of the streamwise velocity component at different y^+. All quantities normalised in wall units. -□-, P180; -○-, P360; -●-, P720; -◇-, P1500; -△-, P2880. (a,c) $\kappa_z \phi_{zz}(\lambda_z)$; (b,d) $\kappa_\varphi \phi_{zz}(\lambda_\varphi)$. (a,b) $y^+ = 5$; (a,b) $y^+ = 15$; (c,d) $y^+ = 30$; (c,d) $y^+ = 60$. The dashed lines indicate the location of the spectral peak at $y^+ \approx 15$: $\lambda_z^+ \approx 1000$, $\lambda_\varphi^+ \approx 120$.

same wall-normal locations in bulk units. From the κ^{-1} shape at the upper part of the logarithmic layer ($y/R \approx 0.2, 0.4$) the large Reynolds number streamwise spectra ($\mathrm{Re}_\tau \geq 720$) shift towards single-peak spectra in the bulk flow region, while they decay in magnitude. The spanwise spectra become less relevant, when entering the region near the pipe axis, which is due to the cylindrical coordinate system.

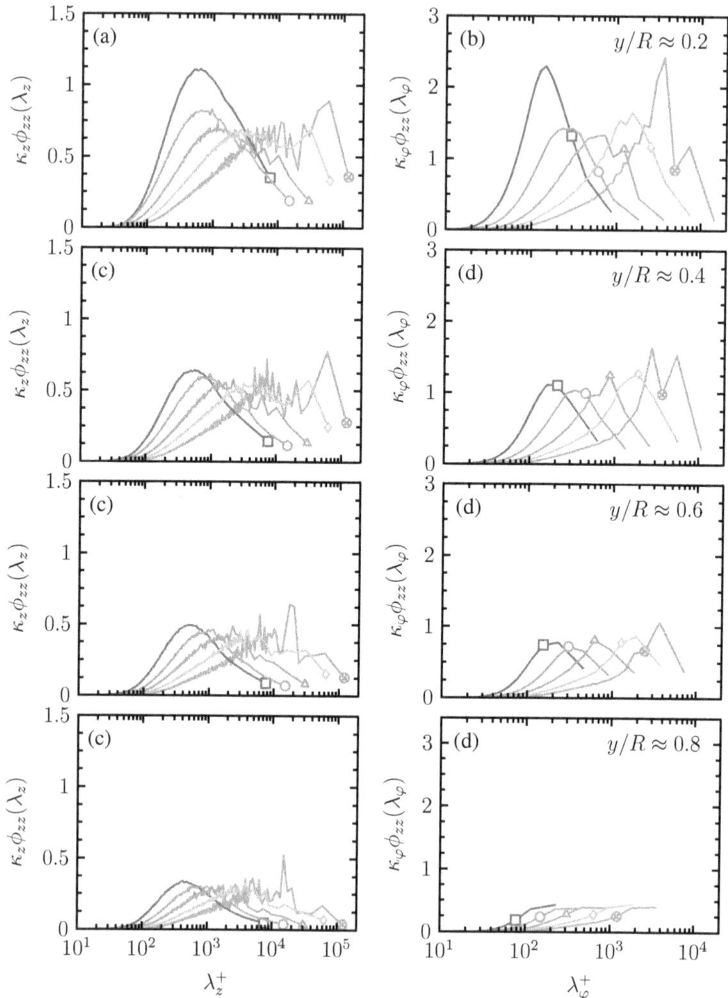

Figure E.2: One-dimensional pre-multiplied energy spectra of the streamwise velocity component at different y/R. All quantities normalised in wall units. -□-, P180; -○-, P360; -●-, P720; -◇-, P1500; -△-, P2880. (a,c) $\kappa_z \phi_{zz}(\lambda_z)$; (b,d) $\kappa_\varphi \phi_{zz}(\lambda_\varphi)$. (a,b) $y^+ = 5$; (a,b) $y^+ = 15$; (c,d) $y^+ = 30$; (c,d) $y^+ = 60$.

Appendix F

Supplementary Figures to Chapter 5

As an appendix to chapter 5, figures F.1 and F.1 show instantaneous iso-contours of unfiltered streamwise velocity fluctuations, filtered streamwise velocity fluctuations, and the inter-scale energy flux term.

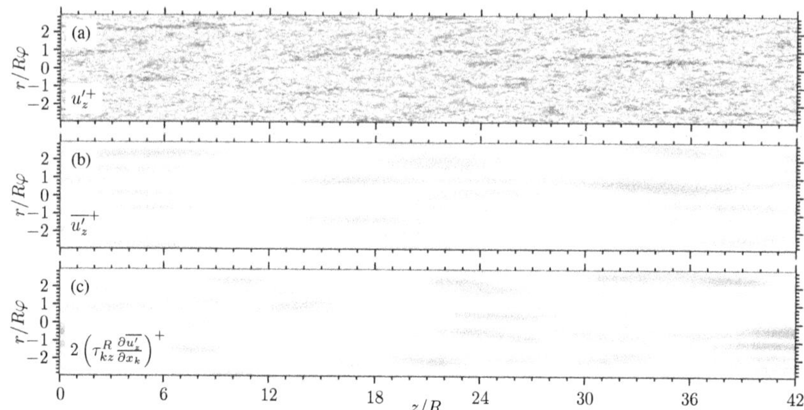

Figure F.1: Iso-contours in a wall-parallel plane at $y^+ = 100$. (a) Unfiltered streamwise velocity fluctuation u_z'. (b) Low-pass filtered streamwise velocity fluctuation $\overline{u_z'}$. (c) Instantaneous inter-scale energy flux term, $2(\tau_{kz}^R \partial \overline{u_z'} / \partial x_k)^+$. All terms normalised in wall units. Filter lengths: $\lambda_z^+ = 9000$, $\lambda_\varphi^+ = 1000$. (a,b) Values ranging from -5 (dark blue) to 5 (dark red). (c) Values ranging from -0.03 (dark blue) to 0.03 (dark red).

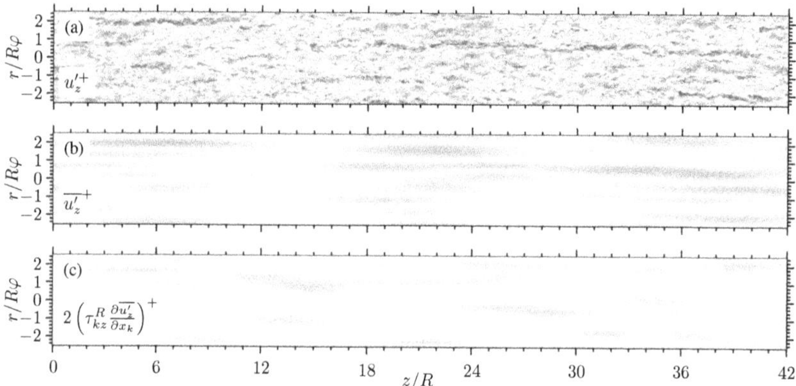

Figure F.2: Iso-contours in a wall-parallel plane at $y^+ = 300$. (a) Unfiltered streamwise velocity fluctuation u_z'. (b) Low-pass filtered streamwise velocity fluctuation $\overline{u_z'}$. (c) Instantaneous inter-scale energy flux term, $2(\tau_{kz}^R \partial \overline{u_z'} / \partial x_k)^+$. All terms normalised in wall units. Filter lengths: $\lambda_z^+ = 9000$, $\lambda_\varphi^+ = 1000$. (a,b) Values ranging from -5 (dark blue) to 5 (dark red). (c) Values ranging from -0.03 (dark blue) to 0.03 (dark red).

www.ingramcontent.com/pod-product-compliance
Ingram Content Group UK Ltd.
Pitfield, Milton Keynes, MK11 3LW, UK
UKHW021652190726
13853UKWH00001B/213

9 783736 973695